Rainer Müller, Franziska Greinert
Quantum Technologies

Also of Interest

Quantum Information Theory. Concepts and Methods
Joseph M. Renes, 2022
ISBN 978-3-11-057024-3, e-ISBN (PDF) 978-3-11-057025-0,
e-ISBN (EPUB) 978-3-11-057032-8

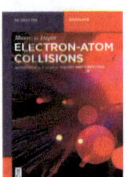

Electron–Atom Collisions. Quantum-Relativistic Theory and Exercises
Maurizio Dapor, 2022
ISBN 978-3-11-067535-1, e-ISBN (PDF) 978-3-11-067537-5,
e-ISBN (EPUB) 978-3-11-067541-2

*Quantum Mechanics. An Introduction to the Physical Background and
Mathematical Structure*
Gregory L. Naber, 2021
ISBN 978-3-11-075161-1, e-ISBN (PDF) 978-3-11-075194-9,
e-ISBN (EPUB) 978-3-11-075204-5

Polarization of Light. In Classical, Quantum, and Nonlinear Optics
Maria Chekhova, Peter Banzer, 2021
ISBN 978-3-11-066801-8, e-ISBN (PDF) 978-3-11-066802-5,
e-ISBN (EPUB) 978-3-11-060509-9

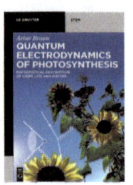

*Quantum Electrodynamics of Photosynthesis. Mathematical Description of Light,
Life and Matter*
Artur Braun, 2020
ISBN 978-3-11-062692-6, e-ISBN (PDF) 978-3-11-062994-1,
e-ISBN (EPUB) 978-3-11-062700-8

Rainer Müller, Franziska Greinert

Quantum Technologies

For Engineers

DE GRUYTER

Authors

Prof. Dr. Rainer Müller
Physik und Physikdidaktik
Technische Universität Braunschweig
Bienroder Weg 82
38106 Braunschweig
Deutschland
rainer.mueller@tu-braunschweig.de

Franziska Greinert
Physik und Physikdidaktik
Technische Universität Braunschweig
Bienroder Weg 82
38106 Braunschweig
Deutschland
f.greinert@tu-braunschweig.de

ISBN 978-3-11-071744-0
e-ISBN (PDF) 978-3-11-071745-7
e-ISBN (EPUB) 978-3-11-071750-1

Library of Congress Control Number: 2023944099

Bibliographic information published by the Deutsche Nationalbibliothek
The Deutsche Nationalbibliothek lists this publication in the Deutsche Nationalbibliografie;
detailed bibliographic data are available on the Internet at http://dnb.dnb.de.

© 2024 Walter de Gruyter GmbH, Berlin/Boston
Cover image: Quardia / iStock / Getty Images Plus
Typesetting: VTeX UAB, Lithuania
Printing and binding: CPI books GmbH, Leck

www.degruyter.com

Preface

The emerging field of quantum technologies has attracted much attention in recent years. News portals regularly report on the latest developments in quantum computing and quantum simulations. Quantum communication and quantum sensors have technological potential for direct application. Large international projects, large companies, and small startups are driving the development of quantum technologies. The potential of quantum technologies for disruptive innovation is a frequently discussed topic.

What is it about quantum technologies that is so new and unique? It is not quantum physics, the underlying physical theory. Quantum physics has been well established and excellently confirmed for 100 years. All of the physics that is now being used in quantum technologies has always been part of it – it was just never really believed. Superposition states and entanglement, Schrödinger's cat, Bell's inequality, and the quantum mechanical measurement process seemed too strange. What is new about quantum technologies is that all these phenomena, long considered more of a playground for philosophy, are now being taken seriously, put to use, and turned into concrete technological products.

Transferring subtle quantum effects from the laboratory into a working product is a formidable technological challenge. No other technological device requires such a detailed level of control as a quantum computer. Its crucial components, the qubits, need to be controlled and managed down to their atomic states in order for a quantum computer to work. To make this possible and then to unlock the broad potential of its applications – from cryptographic techniques to pharmaceutics and the search for new, climate-friendly synthesis methods in chemistry – requires people with very different backgrounds: physics, engineering, computer science, and chemistry.

New study programs focusing on quantum technologies are emerging in many places – often with a special emphasis on engineering. Students in these programs come from very different backgrounds and with very different prior knowledge. As future developers and users of quantum technologies, they need to acquire a broad foundation of interdisciplinary knowledge. This is the audience for which this book is intended.

Quantum technologies require a new approach to quantum physics that emphasizes different aspects than has traditionally been the case. The focus is no longer on the historically important contexts such as atomic physics but on the aforementioned "peculiarities" of quantum physics that are exploited in quantum technologies. Understanding them is intellectually challenging but by no means impossible. The "basic rules of quantum physics", which we present in Chapter 3, allow an intuitive understanding of the description of quantum phenomena. They serve as reasoning tools and help to develop a "gut feeling" for quantum effects.

A certain amount of mathematization is unavoidable to understand the subject. We have tried to keep it to a minimum and to avoid technical jargon and confusing notation. We have always used the simplest nontrivial examples for illustration. The numerous exercises are intended to help the reader understand concrete arguments step by step.

https://doi.org/10.1515/9783110717457-201

The guiding principle in the selection of topics was not to present as much as possible, but as little as possible.

The book is structured as follows: The first chapter gives an overview of quantum technologies and the current state of development of promising applications. The second chapter briefly reviews the physics needed to understand the underlying hardware of qubits and quantum sensors. The actual discussion of quantum physics then begins in Chapter 3. Using two key experiments – the double-slit experiment and the anticoincidence experiment of Grangier, Roger, and Aspect – the formalism of quantum physics is introduced. Topics like superposition, entanglement, and the quantum mechanical measurement process, which are relevant to quantum technologies, are discussed in detail.

Chapters 4–6 are devoted to the various fields of quantum technologies. Here, the operation of quantum sensors, the fundamentals of quantum communication and quantum cryptography, and the use of quantum algorithms in quantum computers are explained with numerous examples and practical applications.

How to use this book

This book is intended to be used as the basis for a lecture or for self-study. If you work through it step by step, you should try to solve the examples yourself before following the solution. Chapters 2 and 3 provide the (physical) basics and the vocabulary for understanding the later chapters on quantum technologies.

However, you may have already heard a traditional quantum physics lecture or simply not be interested in understanding these details. It is also possible to skip chapters 2 and 3 and – after the introductory overview in chapter 1 – go directly to one of the chapters on quantum sensing (4), quantum communication (5), or quantum computing (6), perhaps skipping the examples and equations and just reading the main text quickly. This kind of approach is also possible and is a good way to get an overview of the topic. Of course, you can always return to the basic chapters if you find that you are interested in a particular quantum phenomenon or want to understand the background. We recommend that you read at least section 3.2 on the basic rules of quantum physics, where the fundamental quantum phenomena are discussed in a qualitative way.

Braunschweig, Germany   Rainer Müller and Franziska Greinert
October 2022 and July 2023

Contents

1 Quantum technologies: An overview

Modern quantum technologies exploit the fundamental characteristics of quantum physics for technical applications. It is a young and rapidly developing field that has only recently reached the level of public attention it enjoys today. The advent of quantum technologies marks a new stage in the development of quantum physics, which itself is more than 100 years old and well researched.

In quantum technologies, genuine quantum phenomena like entanglement or the existence of superposition states, which were previously of purely scientific interest, are now being made technologically useful. They move from basic research to engineering applications. Instead of showing up only a few times in fragile laboratory setups, these quantum effects must now be reliably applied in usable products. Concepts such as *Technology Readiness Level* (TRL) and aspects of economic utilization are coming into focus. The field is considered a disruptive key technology, with high expectations for future economic impact.

Remarkably, quantum technologies utilize the same effects that Einstein, Bohr, and Schrödinger discussed in 1935 out of purely scientific and epistemological interest. The formalism and the basic structure of quantum physics have not changed since then. What has changed fundamentally are the experimental capabilities and the mindset of the players – the willingness to utilize the potential of quantum physics for fundamentally new technological approaches.

The year 2018 marks a visible starting point for quantum technologies. Two major initiatives, the Quantum Technology Flagship in the European Union and the National Quantum Initiative in the United States, were launched at this time, giving the field a tremendous boost. Comparable initiatives also exist in countries like the United Kingdom, China, or Canada. The pillar model of the Quantum Flagship (Fig. 1.1), which structures the content of quantum technologies, has become generally accepted. We therefore base our overview on it. It divides the field of quantum technologies into four pillars: communication, computation, simulation and sensing/metrology. They are grounded in basic science, and interwoven with transversal structures (engineering/control, software/theory, education/training).

1.1 Quantum computing

Quantum computing is arguably the most spectacular of the quantum technologies. The field is receiving a great deal of public attention due to the potential of quantum computers to solve a limited class of problems significantly faster than classical computers will ever be able to. Although the technical development of quantum computers into operational devices is still in its early stages and, for many industries, the applicability of quantum algorithms to practical problems is still under investigation, an enormous economic impact of quantum computers is foreseeable.

https://doi.org/10.1515/9783110717457-001

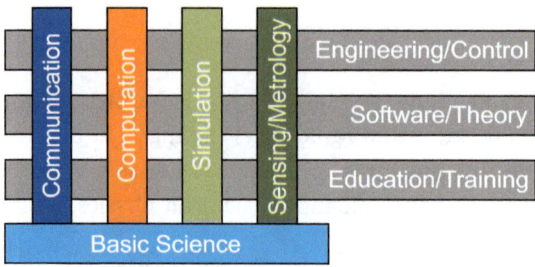

Figure 1.1: Outline of quantum technologies (pillar model of the European Quantum Flagship).

The *qubit* is the central concept in quantum computing. It generalizes the bit, the basic unit of information in classical computer science. A classical bit can take on two values (0 or 1) that are mutually exclusive. The qubit generalizes this concept. It uses the superposition states of quantum physics, which do not occur in classical physics. A qubit can exist not only in the states 0 and 1 but also in arbitrary superpositions of both.

A quantum mechanical superposition of 0 and 1 does not mean a return to analog values such as 0.6. In a superposition state of the qubit, both individual states are actually represented and can, for example, be made to interfere. This is the central point in which quantum computing differs from classical computing. We will discuss the character of superposition states in detail in the following chapters.

The existence of superposition states introduces a special kind of *parallelism* into the computation. There are no if/then/else constructions in quantum algorithms, and there is no need to query for 0 or 1. The two alternatives are processed in parallel since both are represented in the superposition state. However, there is one crucial limitation to this "parallelism". Due to the peculiarities of the quantum mechanical measurement process, a single measurement can never read out all the information contained in the state. In particular, one measurement cannot separately read out the results for several parallel "branches" of the superposition.

The quantum mechanical state of the qubits contains all the information obtained in the parallel computation – but it cannot be fully retrieved. Like with the fairy in the fairy tale, one has only one free question during the measurement. This circumstance makes the construction of *quantum algorithms* very difficult. The challenge is to formulate that one question so that the answer contains the desired solution to the particular problem.

Quantum algorithms

The large general interest in quantum computing began in 1994 with the then quite unexpected discovery by Peter Shor that quantum computers could perform a practically relevant task – the factoring of large numbers – exponentially faster than classical algorithms. This task is practically relevant because many of the current encryption methods are based on the fact that factoring sufficiently large numbers is an enormously difficult problem for classical computers that cannot be solved in a reasonable amount of time. With a working quantum computer, classical encryption methods could possibly be broken.

After Shor's algorithm had identified a first class of problems in which quantum computers could, in principle, be superior to classical computers, the search began for other quantum algorithms that offered such an advantage. It turned out that the range of problems that could be solved more efficiently by quantum computers seems to be limited. The site quantumalgorithmzoo.org lists about 65 quantum algorithms that are demonstrably or supposedly more efficient than any classical algorithm – most of which, however, are of little practical interest. Besides Shor's algorithm, the most important quantum algorithms are *Grover's algorithm* for searching unordered databases, and various algorithms for solving problems from linear algebra. The latter can be subsumed under the category *singular value decomposition* of matrices – a potentially very important application since an enormous number of modeling tasks from all areas of science and technology are based on matrix operations. Furthermore, quantum computers could efficiently solve *optimization problems*, which are important in industries like chemistry, pharmaceutics, finance and logistics. Hopes are also high for the field of *Quantum Machine Learning*, where quantum algorithms assist in pattern recognition.

More recently, interest in quantum algorithms has moved increasingly towards applicability – both in terms of identifying practical use cases in the various sectors of business and industry, and in terms of realistically implementing the algorithms on the imperfect and highly noisy quantum computers that can be practically built. These *NISQ computers* (*noisy intermediate-scale quantum computers*) will prevail in the first phase of quantum computing. The *Quantum Approximate Optimization Algorithm* (QAOA) is an example of a quantum algorithm suitable for NISQ. However, the long-term goal is to get the unavoidable noise problem under control by quantum error correction methods.

Hardware for quantum computers

For classical computers, there are two fundamentally different architectures: the analog computer, which is based on continuously varying mechanical or electrical quantities (it died out sometime in the 1970s) and the digital computer, which is based on bits and standardized gate operations (such as NOT, AND and OR). Both variants also exist in quantum computing. In contrast to classical computers, the analog variant is seen as having promising development prospects here, especially in the field of quantum simulations.

The quantum algorithms mentioned in the previous section are based on the universal, gate-based quantum computer concept. In quantum computing, the gates and their operations have one crucial difference from classical computing: Everything must be *reversible*. Any irreversibility would destroy the fragile quantum mechanical superposition states on which quantum computing is based. David Deutsch developed the basic principle of the universal reversible quantum computer in 1985. He was able to build on

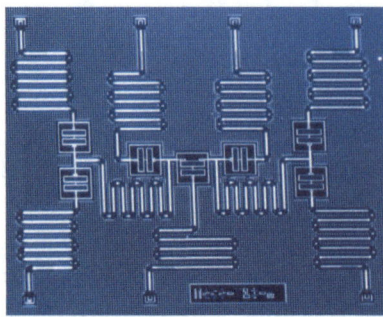

Figure 1.2: Circuitry for seven superconducting qubits in an IBM quantum computer.

Figure 1.3: Cryostat of a quantum computer with superconducting qubits.

earlier work in the theory of reversible computation by Fredkin, Toffoli, Bennett, and others. They had introduced special gates suitable for the purpose, such as the CNOT gate (cf. Chapter 6).

Avoiding irreversibility is one of the crucial experimental challenges in the design of quantum computers. In the physical systems used for the realization of qubits, the respective states must be extremely well shielded against uncontrolled external influences. Nevertheless, it must be possible to address and manipulate them reliably to perform gate operations with them. The physical qubits must also be manufacturable in a controlled and scalable manner. Several physically quite different systems are considered to have the potential to meet these requirements:

1. *Superconducting Qubits*: This is one of today's prevailing technologies for implementing qubits. The hardware is based on superconducting Josephson junctions, capacitors and resonators (Fig. 1.2). For superconductivity to occur, low temperatures are necessary (Fig. 1.3).
2. *Trapped ions*: Ions caught in traps can be manipulated in a controlled way by lasers or microwaves. In more recent approaches aimed at miniaturization and scalability, the ions are trapped on the surface of a chip (Fig. 1.4).

Figure 1.4: Ion trap on a chip. The ions are held in place by static and oscillating electric fields and can be moved spatially to perform computational operations.

3. *Neutral atoms in optical potentials*: Lasers can be used to create spatial "lattice potentials" in which neutral atoms are trapped (in particular atoms in highly excited states, so-called *Rydberg atoms*). With this architecture, a large number of qubits can be realized simultaneously, although currently with high error rates.
4. *Quantum dots in semiconductors*: Quantum dots are intentionally created impurities in semiconductors in which electrons have discrete energy levels. They can be regarded as "artificial atoms". The spin degree of freedom of the electrons is often used to realize this type of qubit because longer coherence times can be achieved in this way. An attractive feature of the semiconductor-based approach is that it allows the use of well-established methods to fabricate semiconductors. In addition, purely electrical control is possible.
5. *Photonic qubits*: In this approach, the quantum states of light are used to process information. To do this, individual photons have to be systematically generated, manipulated, and made to interact. In addition to the gate-based approach to quantum computing, the measurement-based approach is being pursued. Here, sequences of beamsplitters and phase shifters are used to generate highly entangled states of individual photons, on which targeted, problem-specific measurements are then performed.

The *coherence time* specifies how long a qubit state remains usable for quantum computation, i. e., has not been altered by disturbing influences from the environment. **i**

Superconducting qubits and trapped ions are currently the most mature hardware concepts for quantum computing. They are categorized as "application-ready architectures" [1], while the other entries in the list above are classified as "proof-of-performance architectures". In addition, there are even less tested "proof-of-concept architectures", such as topological qubits. When comparing ion-based and superconducting qubits, characteristic differences appear:

1. One notable distinction is the differing possibility for two qubits to interact with each other (connectivity). In superconducting architectures, only neighboring qubits can directly exchange information (as shown in Fig. 1.2). In contrast, in an ion-based computer, all qubits can potentially interact with each other. However, they must be moved around in the trap to do so.

2. Ions have much longer coherence times than superconducting qubits (in the range of several seconds to minutes compared to milliseconds). As a result, they have less noise and lower error rates.

3. Superconducting qubits can perform the individual gate operations much faster (because the ions have to be moved in the trap, which costs time).

Development of quantum computing

The first phase in the development of quantum computing, which lasted about 20 years, was driven primarily by scientific interest. Possible applications were not the focus of interest, but feasibility questions were explored.

Immediately after Shor's algorithm discovery in 1994, the search for physical systems suitable for realizing quantum computers began. Already at the end of 1994, Cirac and Zoller proposed a scheme to realize the CNOT gate using trapped ions. The CNOT gate operates on two qubits. It is important because all conceivable qubit operations can be composed of it, supplemented by some simpler one-qubit operations.

In 1995, Dave Wineland's group at the National Institute of Standards and Technology (NIST) achieved the first experimental implementation of the Cirac-Zoller scheme with a single ^{9}Be$^+$ ion. The basic feasibility of gate operations with qubits was thus demonstrated. Further development for practical technological implementation was not the concern of the scientific community. In basic research, quantum computing was explored in many directions. New quantum algorithms were sought, and quantum error correction schemes were developed to deal with the inevitable coherence losses.

The second phase of quantum computing is characterized by the development towards technological maturity and practical use. It is characterized by the involvement of large companies such as IBM or Google, spearheading the practical development of quantum computers, mainly based on superconducting qubits. The topic received great public attention in 2016 when IBM opened a freely accessible web-based platform, allowing access to real quantum computers for a wide range of users.

With the foreseeable technical feasibility of quantum computers, governments have become active since the mid-2010s. Large funding programs were launched, first in the UK, then in the EU (Quantum Technology Flagship, 2018), the US (National Quantum Initiative, 2018) and various other countries (China, India, Japan). European quantum computer start-ups such as AQT, IQM, or Pasqual have been visible since around 2020.

At present, quantum computing is seen as a disruptive future technology. The predicted applications are in the field of biochemical and pharmaceutical simulations, the optimization of industrial processes, risk and portfolio management in insurance companies and banks, and *quantum machine learning*. Business analysts predict a multi-billion dollar market in the coming decades.

1.2 Quantum simulation

At a conference in 1981, the Nobel Prize laureate Richard Feynman gave a lecture entitled *"Simulating Physics with Computers"*. In his speech, he expressed an insight that can be regarded as either very obvious or very profound, depending on how you look at it. It consists of two parts:
1. Classical systems are not able to simulate larger quantum systems efficiently.
2. Quantum systems are simulated best by other quantum systems.

These two points defined the agenda for the field of *quantum simulation*. It still took several decades before it was taken up and seriously pursued.

Classical computers reach their limits when it comes to fully simulating large quantum systems. This is easy to see just by considering the memory requirements. Even for a moderately large quantum system, it is impossible to store the complete state information on a classical computer. Thus, the classical simulation of quantum mechanical systems already fails before the calculation even begins.

The second part of Feynman's assertion has an obvious plausibility to it. If the state of quantum systems is indeed so complex that any classical description reaches its limits, then why not simulate one quantum mechanical system with another quantum mechanical system? Or, as Feynman himself succinctly puts it [2]:

> ...nature isn't classical, dammit, and if you want to make a simulation of nature, you'd better make it quantum mechanical, and by golly it's a wonderful problem, because it doesn't look so easy.

The basic idea is to model a complex quantum system that is difficult to control with another quantum system that is easier to handle. To do this, one must find a way to map the variables and dynamics of the original system to those of the model system. Thus, by investigating one system, one can infer the properties of the other system. This approach of directly mapping the variables of two systems is called *analog quantum simulation*.

The number of classical bits needed to describe the state of an n-qubit system can be determined by simple counting (cf. p. 77). It scales with 2^n. This is how many complex numbers it takes to store the state of a quantum system. Thus, to fully capture the state of 50 qubits, about $2^{50} \approx 10^{15}$ numbers are needed. For comparison: the Fugaku supercomputer installed in 2020 at the Riken Center for Computational Science in Kobe (Japan) has a storage capacity of 636 TB, i. e., about 0.6×10^{15} bytes.

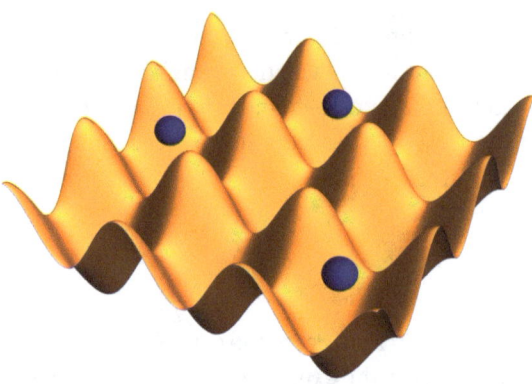

Figure 1.5: Illustration of atoms in a two-dimensional optical lattice potential.

This estimate illustrates why the figure of 50 near-perfect qubits is repeatedly cited as an approximate threshold for quantum supremacy. The 2^n scaling means that adding a 51st qubit doubles the required classical storage capacity – two classical supercomputers are needed. With each qubit added, the required number of classical supercomputers needed to store the state of the system would double.

---

i Of course, Feynman's statement cannot contain the whole truth about the simulation of quantum systems by classical computers. After all, computer simulations of the properties of molecules have long been successfully carried out in theoretical chemistry. Usually, however, approximation methods are used that drastically reduce the information content and focus only on the chemically relevant variables. A well-known example is *density functional theory* (1998 Nobel Prize in chemistry for Walter Kohn), which focuses only on the charge density. The complete state information with all superpositions and entanglements meant by Feynman is not relevant here. However, even density functional theory or comparable approaches quickly reach the limits of computability in simulations of molecules with more than 100 atoms.

For a successful quantum simulation, the model system must be experimentally controlled well enough to meet the following requirements:
1. the quantum state under investigation can be prepared (initialization);
2. the dynamics of the system to be simulated can be reproduced in the model system (ideally, relevant parameters can also be varied);
3. the variables of interest can be read out with reasonable effort.

---

i An example: The investigation of strongly interacting quantum particles in solids is one of the most difficult problems in this field. The interactions can be simulated with atoms in an optical lattice (a periodic potential generated by lasers, cf. Fig. 1.5). In this system, the quantum system can be prepared a well-defined way, and even the variation of the interaction parameters is possible – advantages that are not possible in the original system and that allow a deeper understanding of the original problem. In this way, the study of atoms in optical lattices provides insight into difficult problems in solid-state physics [3].

There are various basic schemes of quantum simulation: from the analog quantum simulation described above to the digital modeling of dynamics in discrete time steps with a gate-based quantum computer. The boundaries between quantum simulation and quantum computing are inherently fluid because a universal quantum computer can always be used for simulations as well.

Quantum annealers

Quantum annealers are a special subgroup of quantum simulators. In materials science, annealing means tempering: a classical procedure of repairing structural defects by heating for a long time. The idea is to bring the system into a state of minimum energy by heating, followed by slow cooling. A quantum annealer comparably solves optimization problems. The problem may be logistical, such as distributing goods on truck trips. The goal is to optimize a utility function that depends on the different distribution possibilities. One of the first experimental applications under real conditions was the bus route planning carried out by Volkswagen using a quantum annealer to optimize traffic flow at the 2019 Web Summit in Lisbon.

In a quantum annealer, the problem to be optimized is described by a specifically adapted coupling between the qubits. The coupling determines the energy of the system and thus defines the utility function. Initially, the qubits are uncoupled. The interaction between them is gradually switched on so that the system always remains in a state of minimum energy. At the end of the switch-on process, there is a high probability that the system will be in a state that represents the desired optimal solution of the problem.

A quantum annealer is not a universal quantum computer. Quantum annealing is an analog optimization method in which, by adjusting the couplings between the qubits, the utility function can be freely programmed within certain limits. As the requirements for the control of the overall system and for the error rates are much lower than in gate-based quantum computing, quantum annealers are technically easier to realize. Therefore, they can offer much higher qubit numbers, although at the cost of less flexibility. In particular, D-Wave's devices attracted early attention with commercial availability and a large number of superconducting qubits (Fig. 1.6).

Use cases for quantum simulation

Quantum simulation and quantum optimization have a variety of applications, the scope of which cannot yet be conclusively foreseen today. Of course, it is obvious to take up Feynman's suggestion and specifically simulate quantum systems. Molecular structures, chemical reaction rates, or – most importantly – the effect of catalysts in chemistry can be studied in this way. Practically, such simulations could be used for drug development (with a large field of application in personalized medicine) or for energy-efficient chemical synthesis with effective catalysts (which could lead to large energy savings, for example, in the production of fertilizers). Other complex quantum systems for which the potential of quantum simulations is being explored are found

Figure 1.6: Processor of a quantum annealer with superconducting qubits.

in solid-state physics and the development of materials with novel properties. There are hopes that the quantum simulations for chemical and materials science questions could also help in combating climate change, for example, in materials for solar cells and batteries or in CO_2 capture and the development of synthetic fuels [4].

Outside of quantum physics, simulations and optimizations are already used in a wide range of application areas: from sequence control in industrial processes and atmospheric models to risk and portfolio management in insurance companies and the financial sector. The extent to which these areas can benefit from quantum simulations, in which fields of application there is quantum superiority at all, and where classical models and heuristic approximation methods may be sufficient after all, will be shown by the current explorations of concrete use cases.

1.3 Quantum sensors

Atomic clocks

Quantum sensing is one of the most mature subfields of the "new" quantum technologies. It has evolved continuously from well-established research in traditional quantum physics. Atomic clocks, for example, have been used to measure time since the 1960s. They are based on the highly precise experimental methods of quantum optics. Since 1967, the second has been defined in the International System of Units (SI) by the transition frequency between two energy levels of the ^{133}Cs atom (which is in the microwave range).

Today's most accurate atomic clocks are based on atomic transitions at optical frequencies. They have an uncertainty of the order of 10^{-18}, that is, one second of deviation in a period greater than the age of the universe. Their continuous evolution is reflected in Fig. 1.7, which shows the four primary atomic clocks of the Physikalisch-

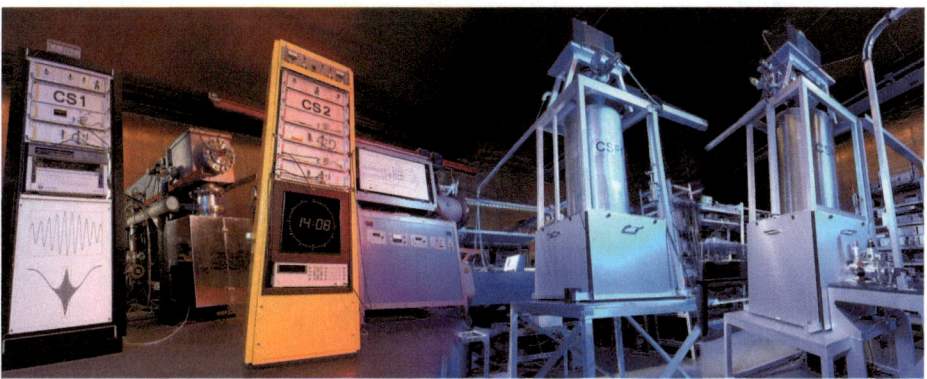

Figure 1.7: The four primary atomic clocks at PTB.

Technische Bundesanstalt (PTB) in Braunschweig, Germany. They were put into operation in 1969 (CS1), 1985 (CS2), 1996 (CSF1), and 2008 (CSF2). The accuracy was increased by four orders of magnitude during this time.

    Atomic clocks are well suited to illustrate the advantages of quantum sensors. Their operating principle is based neither on comparison with an artifact (as in the case of a pendulum clock, which must be calibrated) nor on special material properties (as in the case of a liquid or bimetallic thermometer). The atomic transition that determines the rate of an atomic clock is provided by nature as a universal scale. In the absence of external disturbances, it is the same everywhere and every time. One can take advantage of this and infer the presence of external disturbances from measured deviations from the normal rate. In this way, an atomic clock becomes a quantum sensor.

Gravimetry with atomic clocks

It is precisely the enormous accuracy of atomic clocks that allows them to be used efficiently as sensors. For example, according to Einstein's theory of general relativity, the rate of clocks depends on their height in the Earth's gravitational field. This has immediate consequences for the comparison of atomic clocks at different locations: You have to take their different altitudes into account. Atomic clocks are so accurate that a few years ago, they reached a stage where it was possible to resolve which floor of the same building they were on. Current atomic clocks are even more accurate: they can resolve height differences in the centimeter range. This can be used for practical purposes. When portable atomic clocks become widely available, surveyors will no longer need to determine the elevation of points on the Earth's surface using level rods.

    A closer look reveals additional application areas: Precisely speaking, it is not the height of an atomic clock that determines its rate, but rather the gravitational potential at its location. The gravitational potential depends not only on height but also on other factors, particularly the mass distribution below the atomic clock. In this way, detect-

Figure 1.8: Fluorescence of NV centers.

ing variations in underground density becomes possible; *gravimetry* with atomic clocks becomes possible. They can be used to search for cavities, underground structures, water reservoirs, or natural resource deposits. Also, global mapping of the gravitational potential with satellite-based atomic clocks is being advanced.

The social and technological relevance of these applications is obvious. It is hardly possible to overestimate the usefulness of gravimetry in civil engineering, where it is necessary to locate cavities, underground structures, and wall remnants in the subsurface. Approximately 40 % of all construction sites experience additional costs or delays due to uncertainties in the subsurface [5]. For the UK, one study estimates the cost of finding inadequately documented underground structures in roadworks alone at £5 billion per year [6].

NV centers as quantum sensors

A completely different type of quantum sensor is based on *NV centers* in diamond crystals, a special type of crystal defect. In the diamond crystal, one of the carbon atoms is replaced by a nitrogen atom (N), and an adjacent lattice site remains empty (vacancy V). Due to an additional electron from the surroundings, the defect is negatively charged. In this structure, discrete energy levels emerge in a kind of "artificial atom". These levels can be excited by laser light and show fluorescence: If you illuminate an NV center with a green laser, it emits red fluorescent light (Fig. 1.8).

The energy of the levels in an NV center depends sensitively on external magnetic fields, to a lesser extent also on external electric fields, mechanical stress in the crystal, and temperature. Therefore, NV centers are suitable as sensors for these quantities. For example, the energy difference of two neighboring levels increases with increasing magnetic field (*Zeeman effect*). The corresponding transitions can be excited with microwaves and are visible in the fluorescence spectrum of a simultaneously driven optical transition (Fig. 1.9). With this technique, called *Optically Detected Magnetic Resonance* (ODMR), external magnetic fields can be measured.

NV centers have the same advantage as the atoms in an atomic clock: they are all the same and have universally reproducible properties. An NV-based magnetic field sen-

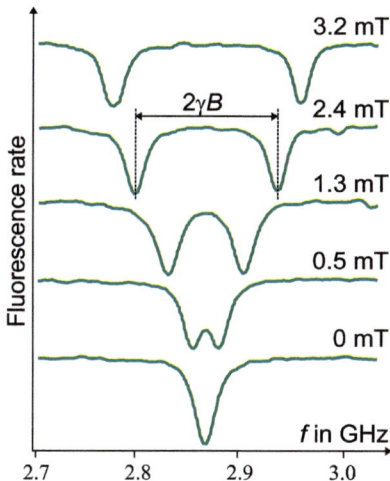

Figure 1.9: The splitting of the energy levels as a function of the magnetic field, which can be read off from the distance between the minima in the fluorescence rate, allows the magnetic field to be measured (data: QZabre).

sor therefore does not require calibration to an external standard. Because their properties do not change over time, NV-based sensors have high reproducibility. Practical difficulties currently remain in the controlled and scalable production of NV centers in diamond crystals and the control of individual NV centers using focused laser light. For these reasons, NV centers are also not the first choice for qubits in a quantum computer.

NV centers are not the only type of highly sensitive magnetic field sensors based on quantum effects. *SQUID sensors* (Superconducting Quantum Interference Device) have been around since the 1960s. They are very sensitive but are based on superconductivity and only work at low temperatures. Elaborate cooling with liquid helium or liquid nitrogen is required; furthermore, miniaturization is complicated by the space requirements of the cryostats needed for cooling.

Sensors based on NV centers do not have these disadvantages because they operate at room temperature. They are small so that high spatial resolution is possible, which is advantageous for medical applications, such as *magnetoencephalography* (MEG) – a medical diagnostic technique that detects the magnetic field of the human brain caused by the electrical activity of nerve cells. This technique is used, for example, to research and diagnose Alzheimer's disease.

Magnetic fields can also be detected with *optically pumped magnetometers* (OPMs), which also work at room temperature. Physically, they are based on atoms contained as a gas in a cell (mostly alkali metals such as potassium, rubidium, or cesium). A laser is used to bring the atoms into a state with certain spin properties. In the presence of a magnetic field, the spin exhibits a field-dependent precession, which can be detected by a change in the absorption rate of the atoms. This allows the magnetic field to be measured (Fig. 1.10).

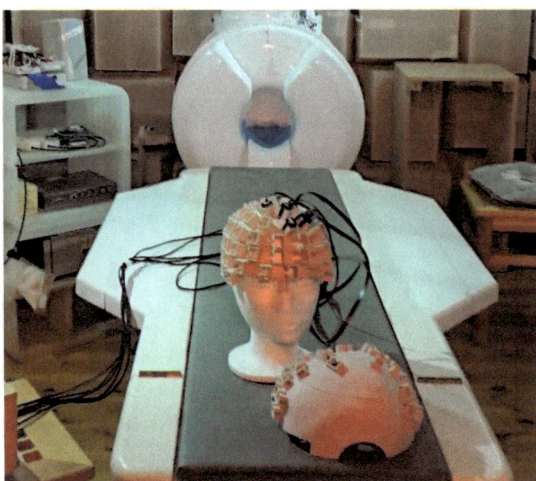

Figure 1.10: Background: SQUID-MEG at PTB Berlin with the reclining area for a person whose head would lie in the MEG device for the measurement. Foreground: two caps with OPM sensors, enabling adaptation to a person's head and movement during the measurement.

Application areas for quantum sensors

Atomic clocks and NV centers are just two examples of quantum sensors. The number of physical systems suitable for measuring environmental variables with quantum effects is immense. By definition, a sensor has to react sensitively to the variable to be measured; the resulting change in its state must be reliably readable. In quantum sensors, quantum effects are used for measurement. However, since the entire physics of semiconductors is fundamentally based on quantum mechanics, it is difficult to draw the line between them and conventional sensors. Typically, nonclassical states or effects (such as superposition states or quantum mechanical phase shifts) are used in quantum sensors to achieve higher sensitivities.

With these higher sensitivities and additional advantages such as miniaturization, reproducibility, robustness, and reliability, quantum sensors are attractive in various fields:

1. *Medicine:* In medical diagnostics, quantum sensors can be used to detect magnetic fields originating from electrical activity in the brain and the heart muscle. This will also allow for the development of new imaging techniques. Furthermore, magnetic field measurements can be used for fast diagnosis of stroke patients.
2. *Biology:* Living cells and microscopic organisms can be imaged without contact or even interaction using imaging optical techniques. Even today, this process is so fast that it is possible to do video recording and dynamic imaging.
3. *Civil engineering:* Sensitive gravimeters can be used to visualize underground structures and hidden cavities and to inspect existing infrastructure.
4. *Earth monitoring and natural resources:* Gravimeters and gradiometers (which measure the gravitational field gradient) allow the detection of groundwater re-

serves and mineral resources or the monitoring of earthquakes and volcanoes. Such sensors can be deployed on the Earth's surface, in satellites, or in aircraft.

5. *Industrial processes and telecommunications:* High-precision time measurement is required in the accurate control of industrial processes and is particularly critical in telecommunications. The challenge is not only to measure time accurately at one location but also to be able to transmit the time information to another location with high accuracy. This requires networks of synchronized clocks.

6. *Detection of particles and gases:* Quantum sensors can be used as very sensitive detectors for gases. The gases are identified by spectroscopic methods. This enables, for example, leak detection in gas pipelines or use in environmental monitoring.

1.4 Quantum communication

Information transmission with light in fiber optic cables was developed as early as the 1970s and has been widely used since the 1990s. In detail, optical data transmission is a complex technology, but the basic principle can be understood simply: Light generated by laser diodes is transmitted in fiber optic cables and detected by photodiodes. The information is encoded by modulating the intensity of the light. Quantum effects hardly play a role in this process. The light used for transmission can be described as a classical electromagnetic wave.

In *quantum communication*, information is transmitted with individual photons. Usually, the information is encoded in their polarization degrees of freedom. Communication with single photons means extremely weak signals. This poses technical challenges. Generating single photons in a controlled and reliable manner is not easy. Sensitive detectors are required for detection; thermal noise is a major problem here.

Controlled *generation of single photons* is an active area of research. Single photons "on demand" are currently technologically infeasible; the inherently statistical character of quantum physics makes this task very difficult. Instead, "heralded single photons" are often used. In a nonlinear crystal (usually barium borate, BBO), an ultraviolet photon is converted into two infrared photons that are highly correlated. This process, called *spontaneous parametric down-conversion*, takes place stochastically. Only one of 10^{11} UV photons is converted into an IR pair. The exact timing of the process is unpredictable and cannot be controlled. The "announcement" of single photons partially compensates for this lack of control over the single process: the two IR photons leave the BBO crystal at different angles – one of them possibly in the direction of a detector. If the detector absorbs this first photon, the detector signal can serve as an "announcement" of the second photon. It is now possible to work with this second photon in a controlled way. Other promising single photon sources are the NV centers discussed in the previous section because they emit exactly one single photon after excitation – but again stochastically in a random direction.

For the *detection of single photons*, avalanche photodiodes (APDs) are commonly used (Fig. 1.11). An APD is the solid-state equivalent of a photomultiplier. It works in a similar way to an ordinary photodiode, where incoming photons lead to the formation of electron-hole pairs. In the APD, the applied reverse voltage is so large that the generated electrons and holes are accelerated, gain energy, and thus trigger an "avalanche" of further electron-hole pairs, which can then be detected as an electrical signal.

Figure 1.11: Avalanche photodiode (APD) used for the detection of single photons.

Quantum cryptography and cybersecurity

Why pursue the approach of communication with single photons at all if the technical hurdles are so high? The answer lies in a fundamental and profound statement about quantum physics: the *no-cloning theorem*, which we will deal with in more detail in chapter 5. It states that it is fundamentally impossible to clone the state of a single quantum object (e. g., a photon). It is impossible to produce an exact copy that is identical in every respect to the original. This statement touches on some fundamental aspects of quantum physics, for example, that a measurement fundamentally changes the state of the measured object, or that one can never read out all the information about the quantum state through a single measurement.

The no-cloning theorem forms the basis of quantum communication. It is especially relevant for *quantum cryptography* – the inherently eavesdrop-proof communication with single photons. Researchers have developed protocols for the secure exchange of cryptographic keys (*quantum key distribution*). With these protocols, it can be ensured that an eavesdropping attempt can be reliably detected due to the disturbance caused by the eavesdropper's measurement of the signal photon. Quantum physics thus provides built-in eavesdropping protection.

After the cryptographic keys, which consist of random sequences of zeros and ones, are reliably distributed, they can be used to encrypt and decrypt the information to be transmitted. An information-theoretically secure transmission that cannot be broken on the transmission path is possible if the keys used are at least as long as the message to be encrypted and are only used once (*one-time pad*).

There are numerous applications because we use encryption daily: Each time a web page is called up with a web browser or smartphone, the transmitted information is encrypted. This is what the "s" (= "secure") in the protocol specification "https" stands for. The transmission of passwords, bank data, and other personal information requires secure encryption. Secure communication is particularly important in areas such as the health system, the financial sector, critical infrastructure (especially energy and water supply), and, of course, in the military and secret service sector.

The encryption methods used today are based on the fact that factorizing large numbers is a difficult task for current computers. As we have seen, this is no longer true for quantum computers. With the advent of powerful quantum computers, the security of the present methods will therefore no longer be guaranteed. This also applies retrospectively: communications intercepted and stored today but not yet decryptable with current resources will become readable as soon as quantum computers master the factorization of large numbers.

Two solution strategies are being followed for this problem: We already mentioned quantum cryptography, where quantum physics is used for secure key exchange. In addition, new encryption methods based on classical information theory are being developed in the field of *post-quantum cryptography*. These methods will no longer be susceptible to attacks from quantum computers.

Quantum teleportation, quantum repeaters, and quantum networks

The development of quantum communication towards maturity follows two main lines: First, to be practical, communication must work reliably over large distances, and second, it is obvious to move from the simple transmitter-receiver model to communication between multiple participants in a *quantum network*.

Communication over greater distances encounters practical difficulties. The photons used in quantum communication are transmitted either in optical fibers or in free space. In both cases, the attenuation of the signal over long distances is a problem – either due to the expansion of the beam in free space or due to absorption in the material of the optical fiber. To keep the latter as low as possible, transmission usually takes place in wavelength ranges with particularly low absorption, the "telecom windows" in the infrared at about 1550 nm. However, even here, optical fibers absorb about 0.2 dB/km, so that after a few tens of kilometers the signal is practically no longer present in the optical fiber.

In free space, the losses are lower so that distances of several thousand kilometers can be bridged with satellite-based communication. Therefore, satellite networks are being pursued in quantum communication, and presumably, they will play an important role in future quantum networks.

Absorption in optical fibers is, of course, also a problem in conventional data transmission with classical light. There, *repeaters* are the solution – a device that receives, processes, and amplifies a signal, then retransmits it. In quantum physics, this runs into difficulties – exactly for the reasons that make quantum communication so secure. The no-cloning theorem prohibits the direct application of the measurement-amplification-retransmission scheme because a single measurement of a single photon can never fully capture its state. This means that the basic prerequisite for retransmitting the information is missing. A repeater based on measurements is fundamentally impossible for individual photons.

With sophisticated methods, *quantum repeaters* are nevertheless possible without violating the no-cloning theorem. In *quantum teleportation*, the quantum state is not copied but shifted spatially. This is done with the help of entangled photon pairs, which have to be shared in advance between sender and receiver. Quantum teleportation is a long-established technique: It was proposed in 1993 [7] and experimentally realized as early as 1997 [8, 9]. In addition to quantum repeaters, temporary memories are also needed for reliable communication (quantum memory). Protocols are being developed that no longer rely on carefully controlled quantum repeaters (trusted repeaters), but instead enable quantum-based end-to-end encryption.

Development in the field of quantum networks is progressing steadily. In China, a 2000 km fibre-optic-based network between Shanghai and Beijing was put into operation in 2017 to exchange quantum cryptographic keys. By 2021, it was expanded to include a 2600 km satellite-based link. In Europe, the EuroQCI initiative was launched to build a secure quantum communications infrastructure spanning the entire EU by 2027.

2 Physical foundations

2.1 Evolution of quantum physics

Historically, quantum physics grew out of the efforts to describe the physical behavior of atoms. Ever since Fraunhofer's observations of the solar spectrum, the *line spectra* of atoms had been known. They arise because gas atoms absorb light only at very specific wavelengths. Light of other, "non-fitting" wavelengths cannot excite the atoms and passes through a gas completely unhindered. The light that is emitted by gas atoms also shows line spectra. The emitted light contains only very specific wavelengths characteristic of each type of atom (Fig. 2.1).

A first, tentative explanation for this behavior was provided by *Niels Bohr's atomic model* of 1913. Bohr postulated that electrons in an atom can only have certain allowed energy values. This feature, which is the lasting innovation of Bohr's model, is called *quantization of energy*. The possible states of the electrons in the atom are called *energy levels*. Line spectra arise from transitions between two energy levels (Fig. 2.2). The frequency of light emitted in a transition between two levels with energy difference ΔE is given by the relation

$$\Delta E = hf, \tag{2.1}$$

where $h = 6.626 \times 10^{-34}$ Js is called *Planck's constant*. Often, the abbreviation $\hbar = h/(2\pi)$ is used, pronounced "h bar".

What remains of Bohr's atomic model is the existence of energy levels in atoms and the model of transitions between them. However, the idea that electrons move around the nucleus on fixed orbits, which is still reproduced in many illustrations today, had to be abandoned in favor of a quantum mechanical description.

Quantum mechanics and atomic physics

Quantum mechanics was developed in the mid-1920s by Schrödinger, Heisenberg, Pauli, Born, and others. It provides a description of atoms, molecules, and solids that is still valid today. With the advent of quantum mechanics, *atomic physics* began to flourish. The structure of atoms, their states and bonds could be unraveled. This achievement helped to put chemistry on a firm theoretical footing.

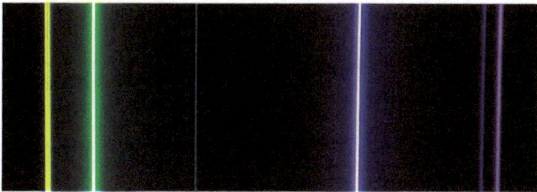

Figure 2.1: Line spectrum of mercury gas.

https://doi.org/10.1515/9783110717457-002

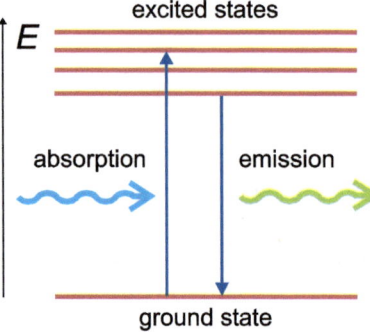

Figure 2.2: Absorption and emission as transition between energy levels.

In *spectroscopy* – the experimental investigation of line spectra – increasingly precise resolution could be achieved. Ever finer effects were observed in the complicated spectra of atoms and molecules and were explained on the basis of quantum mechanics. This detailed understanding of energy levels and the transitions between them is now an indispensable foundation for quantum technologies. The physical realization of qubits on the basis of atoms and ions is based on the complete control of these degrees of freedom.

The mathematical formalism of quantum mechanics can be regarded as a wave theory of matter with some limitations. Electrons are described by a wave function $\psi(x, t)$, and the Schrödinger equation governing their propagation bears some resemblance to a wave equation. However, an all-too-intuitive notion of "electrons as waves" is already made impossible by the fact that ψ is, in general, complex-valued. In addition, the space in which the propagation takes place is not three-dimensional for multi-particle systems. With an increasing number of particles, it quickly becomes very high-dimensional. As we will see, this high dimensionality is ultimately why quantum computers can solve large problems with a relatively small number of qubits.

---

i The term *quantum mechanics* refers to the theory described above, whose basic equation is the Schrödinger equation. It applies to quantum objects with mass, in particular electrons in atoms, molecules, and solids. A more general description is needed for light, where the corresponding quantum objects are called *photons*. They have no mass and travel at the speed of light. Photons can be created or destroyed in absorption and emission processes. They are described by a quantum field theory (quantum electrodynamics), the basic features of which will be discussed later. The broader term *quantum physics* is often used to refer to both of these theories taken together.

Understanding quantum physics

Quantum mechanics is anything but a tangible, easy-to-understand theory. We will have to deal with this fact in the present book. We cannot make it any easier than it is. However, by presenting it from a modern point of view, we can avoid some of the historical baggage that has accompanied the debate about understanding quantum mechanics for decades.

The first problems with the interpretation of quantum mechanics became obvious already during the time of its development, for example, in *Heisenberg's indeterminacy relation* (1927). Here, it became apparent for the first time that something special was happening with certain properties of objects like position and momentum. They are completely unproblematic in classical mechanics but lead to conceptual questions in quantum mechanics.

Although the mathematical formulation of Heisenberg's indeterminacy relation, $\Delta x \times \Delta p \geq h/(4\pi)$, is unambiguous and well-defined, it has been controversial from the beginning about what this means and how to put it into words. Should it be that position and momentum cannot have fixed values at the same time? They cannot be measured simultaneously with arbitrary precision. No simultaneous knowledge of position and momentum is possible? All this has been asserted and written at different times.

Today, the indeterminacy relation can be seen as a first indication that there is indeed a difference between classical and quantum properties: In quantum mechanics, some classically well-defined variables (such as the position or momentum of an electron) actually do not have a definite value in certain quantum states. The corresponding properties cannot be attributed to the quantum object. This is a striking peculiarity of quantum physics and is central to quantum technologies. The whole concept of the qubit is based on the fact that it can be brought into states in which the property "0" or "1" (which is well-defined for classical bits) cannot be ascribed to it. These states are called *superposition states*.

The difficulties raised by the Heisenberg indeterminacy relation were relatively easy to overcome. In the course of further development, however, further problems of interpretation were encountered. Peculiar features of quantum physics were discovered, which made it clear that quantum phenomena cannot be understood using the concepts and views of classical physics. It is precisely these "peculiarities" that are now being exploited in quantum technologies:

1. Max Born's *probability interpretation* (1926), according to which quantum mechanics does not make deterministic predictions about single events (it cannot predict the time of decay of a radioactive atomic nucleus) but only statistical predictions about the measurement results for many repetitions of the same experiment (the decay statistics of many atomic nuclei);
2. the special status of *measurement* in quantum physics: Measurements seemed to elude theoretical description and require separate rules – a difficulty that became known as the "measurement problem of quantum mechanics";
3. the superposition states of macroscopic objects that are predicted by quantum mechanics (Schrödinger's cat, 1935) but could not be reconciled with our everyday experience;
4. the prediction of *entanglement* of multiple quantum objects (Schrödinger, 1935), which manifests itself in nonlocal correlations between measurement results (Einstein, Podolsky, Rosen, 1935; Bohr, 1935);
5. Bell's inequality (1964), which is a statement about local realistic alternative theories to quantum mechanics. After being experimentally verified, it practically ruled out these "hidden-variable theories".

For a long time, the deeper and more serious study of these effects was hampered by the fact that they were attributed less to physics than to philosophy. Niels Bohr's "Copenhagen interpretation" of quantum mechanics was the prevailing interpretation of quantum mechanics for half a century. It focused strongly on epistemology and was formulated rather nebulously in the original papers. Because of Bohr's personal authority, it was widely accepted, though hardly understood.

The general attitude towards the conceptual foundations of quantum mechanics, widespread until the 1980s, is most clearly expressed in the statement: *"I think I can safely say that nobody understands quantum mechanics"*, which Richard Feynman – one of the most important physicists of the 20th century – wrote in his book *"The Character of Physical Law"*. The occupation with the fundamentals of quantum mechanics was not the focus of research. At times, it had something downright disreputable about it. This is illustrated by the anecdote told by Alain. Aspect in his short speech at the 2022 Nobel Prize dinner: When he told John Bell about his plan to verify his inequality experimentally, Bell's first reaction was: "Do you have a permanent position?" In Bell's view, a scientific career could not be built on this topic at that time.

The problems of interpretation, however, always concerned only our understanding of quantum physics, never the theory itself. No experiment in atomic physics has ever failed because the experimenter was unclear about the nature of the quantum measurement process.

The experimental and theoretical advances since the 1990s have led to a new and better understanding of quantum mechanics – without changing the theory itself. Almost all of the "difficulties" listed above, which contributed to the perception that "nobody understands quantum mechanics", are now considered not as "bugs" but as "features" and are being exploited in quantum technologies.

2.2 More recent developments

In the more than 100 years of its development, quantum physics has found an enormous range of applications in science and technology. It has been brilliantly confirmed experimentally, and much of modern physics would be inconceivable without it. Three lines of development are particularly important for quantum technologies.

Solid-state physics

Quantum mechanics also applies to electrons in solids. Since the late 1920s, it has been used to describe electrons in periodic crystals (Bloch, 1928), and the theory of energy bands has been developed (Wilson, 1931). The path to the technical application was through the use of doped semiconductor materials. The development of the transistor (Fig. 2.3) by Shockley, Bardeen, and Brattain in 1947 was the starting point for the tremendous boom in semiconductor technology that characterizes modern technology and society today. Despite its enormous technical and economic importance, and although it is based on genuine quantum mechanical effects, semiconductor technology is not considered one of the new quantum technologies. The reason is that its effects are macroscopic in nature, appearing even when very many charge carriers are involved. In contrast, "second-generation" quantum technologies are based on the more subtle effects that occur in a single or a few isolated quantum objects.

These kinds of quantum effects can be observed in special systems of solid-state physics: in superconducting systems, at impurities in diamond crystals, or in silicon quantum dots. We will discuss these systems, which can be used to construct qubits and quantum sensors, in more detail in later chapters. A major advantage of solid-state sys-

Figure 2.3: The inside of a transistor used in computer systems in the 1960s.

tems in quantum technologies is the mature and technically well-tested semiconductor fabrication methods that can be relied upon for suitable applications.

Quantum optics and photonics

Quantum optics deals with the interaction of light and atoms, particularly with effects in which their quantum nature becomes apparent. Historically, it can be traced back to the beginnings of quantum physics via the molecular beam experiments of Stern and Rabi and the microwave techniques studied to develop radar during World War II. A particular milestone was the invention of the laser (Maiman, 1960), which became an indispensable tool in all quantum optics experiments.

Many experimental approaches and techniques developed in quantum optics are now used as important building blocks in quantum technologies. First and foremost, there are the experimental techniques developed in precision spectroscopy for the increasingly precise measurement of line spectra. They include the trapping of atoms and ions with electromagnetic fields (Paul trap, 1958) and the subsequent laser cooling, in which their velocity is reduced by the interaction with suitably tuned laser light (Fig. 2.4). Today, it is possible to cool atoms, ions, and molecules to temperatures in the microkelvin range. Precisely measured laser or microwave pulses allow controlled state manipulation – essential in quantum technologies for controlling qubits.

These techniques illustrate a characteristic of quantum optics: the development towards ever higher precision, accompanied by ever better control of the quantum objects. Quantum optics is a very "clean" discipline. It studies fundamentally simple, theoretically well-understood systems. This high level of control is essential for the development of quantum technologies.

A related branch of quantum optics deals with the properties of light in all its facets. *Photonics* is concerned with the transmission of information by light, especially in optical fibers. This requires tailor-made sources, amplifiers, modulators, and detectors. The application-oriented field of photonics is now considered one of the *enabling tech-*

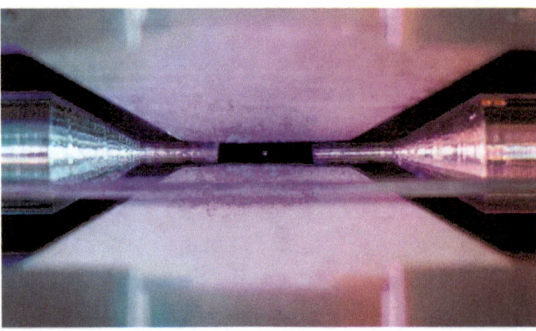

Figure 2.4: The small dot in the center shows the fluorescence light emitted by a single strontium ion in an ion trap. The distance between the two electrode tips is about 2 mm.

nologies for quantum technologies. Here, the generation, transmission, and detection of single photons play an important role, especially in quantum communication. When transmitting information with light, the polarization degrees of freedom are often used to encode information.

Identifying genuine quantum properties of light is surprisingly difficult. It was not until 1979 that Walls and Milburn published the article *"Evidence for the Quantum Nature of Light"*, in which they demonstrated the first true experimental evidence for photons: the antibunching effect, which describes the statistics of photons emitted by a single photon emitter. On the other hand, the "school example" of the photoelectric effect is not considered a true quantum effect because it could already be explained in the 1920s within the framework of a semiclassical theory (quantized atoms interacting with a classical electromagnetic field).

A modern understanding of quantum mechanics

Since the 1990s, a more modern view of quantum mechanics has emerged. It is not the theory itself that has changed but our understanding of its more peculiar features. Both experimental and theoretical contributions have played a role.

Experiments with single quantum objects, which had only been debated for decades, now became feasible in reality. They moved from the realm of mere thought experiments to that of experimentally tangible reality. Examples are the double-slit experiments, in which whole atoms were made to interfere one by one (Mlynek et al., 1991), the anticorrelation experiments with single photons by Grangier, Roger and Aspect (1986) – we will discuss them in chapter 3 – or the experimental tests of Bell's inequality (Aspect et al., 1982). The fact that each of these experiments brilliantly confirmed the often nonintuitive predictions of quantum physics contributed significantly to the acceptance of its truly nonclassical character.

On the other hand, there were theoretical developments that gave new impetus to the stagnating interpretation debate. In particular, the theory of decoherence (Zeh, 1970;

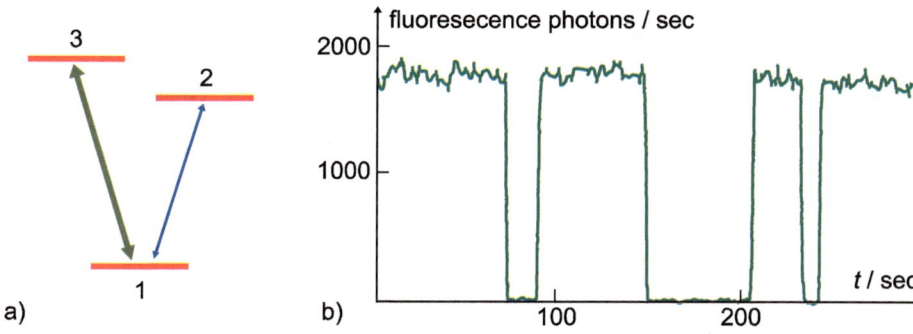

Figure 2.5: Atomic transitions and counting rate of fluorescence photons in the "Quantum Jump" experiment [10].

Zurek, 1991) shed new light on the transition between quantum and classical physics and on the quantum measurement process. This approach – which we will also discuss in more detail below – removes much of the mystery from the problem of Schrödinger's cat and the quantum measurement process. The argument: In the traditional discussion of these issues, oversimplified theoretical models were considered which neglected the influence of the ever-present environment on a quantum system. If the influence of the environment is taken into account, the observed nonappearance of superposition states for macroscopic objects can be explained. The modern interpretation of quantum mechanics is based on the theory of decoherence and, unlike the Copenhagen interpretation, largely dispenses with philosophical issues. It is often called the "minimal interpretation" of quantum mechanics and poses few problems of interpretation. Further advances in quantum technology will most likely lead to a habituation effect: We will get used to the strange phenomena of quantum physics as they are utilized in all kinds of technical devices.

As an example of the experimental progress that has led to a better conceptual understanding of quantum physics, we consider the "quantum jump" experiments that were conducted independently in several laboratories in 1986. Their possible outcome was not uncontroversial in advance.

The experiments were conducted with single ions in ion traps. Three energy states are relevant, shown schematically in Fig. 2.5(a). The ion has a ground state 1 and a long-lived excited state 2, which can be excited by resonant laser radiation, but only with low probability (thin blue arrow). The ion can be in the ground state 1, in the metastable state 2, or in a superposition of both. In the experiment, a measurement was performed to determine the state of the ion at a given time.

To "test" the state of the ion, another transition was used between state 1 and a third state 3. This transition is much "faster" than the first one (indicated by the thick green arrow). When resonant laser light hits the ion in the ground state 1, it undergoes a rapid succession of transitions between states 1 and 3, emitting intense visible fluorescent light. On the other hand, if the laser light hits the ion in the metastable state 2, no fluorescence light is emitted because the incident laser light is not resonant with any of the possible transitions. Thus, the detection of fluorescence light means: the ion is in ground state 1; the absence of fluorescence light means: the ion is in metastable state 2. The question to be answered in the experiment was: What happens if the ion is neither in state 1 nor in state 2, but in a superposition of both?

Fig. 2.5(b) shows the result obtained in the experiment of Dehmelt et al. [10]. In this experiment, a single Ba$^+$ ion in an ion trap was continuously irradiated with light at both transition frequencies. For practical reasons, the excitation in state 2 was not direct, but via another state, which is irrelevant in our context. The intensity of the emitted fluorescence light was registered by a detector. It is shown in the right part of the figure as a function of time. We recognize periods of bright fluorescence interrupted by periods of darkness. The duration of the phases is considerable, lasting 30 or more seconds. The transition between the two phases is abrupt. It is unpredictable when exactly a transition will occur; the duration of the phases follows a statistical distribution.

It is noteworthy that the fluorescence light is either "bright" or "dark". Intermediate values of fluorescence intensity are not found. From time to time, excitation from state 1 to state 2 (or the reverse transition) occurs. This is manifested by the sudden onset or interruption of fluorescence light. The transition is not gradual, but happens from one instant to another: *quantum jumps* are being observed. A striking feature of this experiment is that the emitted fluorescent light can be seen with the naked eye. The change between bright and dark phases caused by quantum jumps of a single ion is directly perceptible.

Today, the "Quantum Jump" method has found practical applications in quantum technologies. It is used, for example, in ion-based quantum computers to read out qubits at the end of a calculation.

2.3 Atoms and their spectra

Atomic energy levels

The structure of atomic *energy levels* is often complicated. For a general understanding, it is helpful to start with the simplest of all elements: hydrogen with only one electron. On the basis of the structure found there, the energy levels of higher atoms can be explained.

Mathematically, the possible states of the electron in the hydrogen atom are obtained by solving the *Schrödinger equation* for the wave function $\psi(x, t)$. Similar to the phenomenon of standing waves, where only certain wavelengths occur in a closed space, e. g., an organ pipe, there is a discrete set of possible electron states in the atom. One of them is shown in Fig. 2.6(b). According to the probability interpretation of the wave function, $|\psi(x, t)|^2$ gives the probability of finding the electron at location x when a position measurement is made. In Fig. 2.6(b), the probability is represented by the brightness of the "electron cloud". The states of the electron in the hydrogen atom are characterized by integer or half-integer *quantum numbers*:

1. the principal quantum number n, which gives the main contribution to the energy of the state via the formula:

$$E_n = -\frac{m_e e^4}{8h^2\epsilon_0^2}\frac{1}{n^2};$$

(2.2)

2. the quantum number ℓ, which can take on integer values between 0 and $n - 1$ and characterizes the orbital angular momentum of the electron;

3. the quantum number m, which can take integer values from $-\ell$ to $+\ell$ (i. e., a total of $2\ell + 1$ values). m is called *magnetic quantum number* because it is relevant for the energy shifts in magnetic fields;

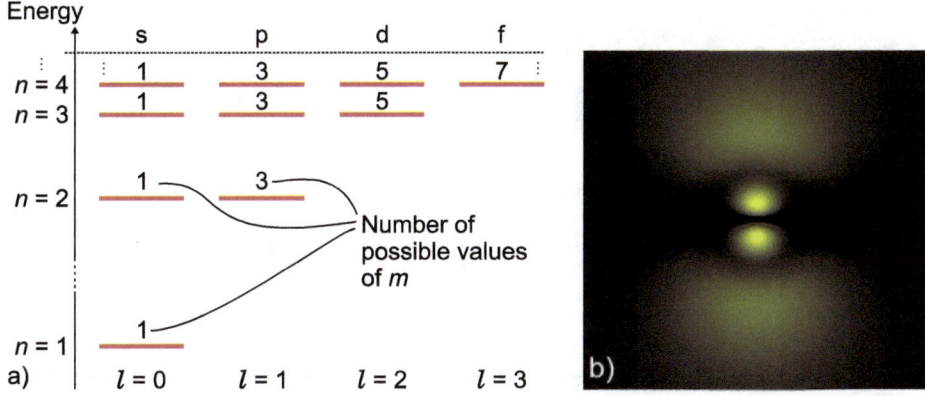

Figure 2.6: (a) Term diagram of the hydrogen atom, (b) probability distribution for the state with $n = 3$, $\ell = 1$ and $m = 0$.

4. the spin quantum number s with two possible values $\pm\frac{1}{2}$, which gives the intrinsic angular momentum (spin) of the electron in units of $\hbar$.

The left side of Fig. 2.6 shows the *term diagram* of the hydrogen atom: a representation of the possible states and their energies. On the vertical axis, the energy is plotted according to Eq. (2.2). The ground state ($n = 1$) has an energy of –13.6 eV. Transitions between states with different values of n typically lie in the optical or infrared region and are therefore excitable with lasers.

Fine structure and hyperfine structure

To the right, the various angular momentum quantum numbers ℓ are plotted in the term diagram. For historical reasons, the different values of ℓ are denoted by the letters s, p, d, f. The number of possible values of m (i. e., $2\ell + 1$) indicates the number of states hidden behind each of the levels. All states with different values of ℓ and m have the same energy – at least at this level of resolution.

This "degeneracy" of energy levels is removed when we consider more subtle effects: An orbital or spin angular momentum is always accompanied by a magnetic moment, which leads to various interactions: the interaction between the electron's orbital and spin angular momentum, the interaction between the angular momentum of the electron and the angular momentum of the nucleus, and relativistic and quantum electrodynamics corrections.

All these effects lead to small energy shifts of the individual states, depending on the angular momentum quantum numbers. The individual levels in Fig. 2.6 are therefore split with respect to energy. These energy shifts are much smaller than the energy differences described by the principal quantum number and are therefore called *fine structure* and *hyperfine structure* of the energy levels. Typical transition frequencies are on the order of GHz, in the microwave or radio frequency range.

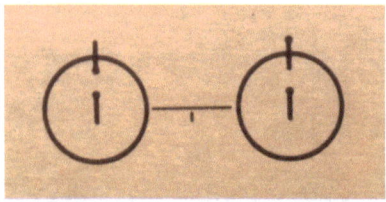

Figure 2.7: Detail of the Pioneer plaque. The image illustrates the two possible settings of electron and nuclear spin and the associated energy splitting; the small vertical line symbolizes the length standard defined by the 21-cm line.

Example: The ground state of the hydrogen is split by the interaction between the spins of the electron and the atomic nucleus. The transition frequency of this hyperfine splitting is at 1420 MHz. Give the wavelength of the emitted electromagnetic radiation and the photon energy in eV.

Solution: We calculate the wavelength λ from the frequency f with the relation $\lambda f = c$, which is valid for all kind of waves. c is the propagation speed of the wave, here the speed of light. We obtain:

$$\lambda = \frac{c}{f} = \frac{3 \times 10^8 \, \frac{m}{s}}{1420 \times 10^6 \, \frac{1}{s}} = 21.1 \text{ cm.} \tag{2.3}$$

The radiation emitted during the transition has a wavelength of 21 cm and is thus in the microwave range. With $E = hf$ and the definition of the unit eV 1 eV $= 1.602 \times 10^{-19}$ J we find 5.9×10^{-6} eV for the energy of a photon of this radiation.

Due to the ubiquity of hydrogen in space, the 21-cm line of hydrogen is the most common spectral line in radio astronomy. On the famous Pioneer plaque (Fig. 2.7), which left the solar system with the Pioneer 10 and 11 spacecraft, the 21-cm line was therefore used as a length standard to specify the size of humans and the position of the sun in space.

The fact that the hyperfine structure of atoms can be used as a time standard is exploited in atomic clocks. The satellites of the European Galileo navigation system are equipped with atomic clocks based on the 21-cm line of hydrogen. In the International System of Units (SI), the definition of the second is based on a hyperfine transition in cesium. The 1967 definition of the second reads as follows: "The second is the duration of 9 192 631 770 periods of the radiation corresponding to the transition between the two hyperfine levels of the ground state of the cesium 133 atom".

Qubit realizations with atoms and ions
Successful qubit implementations have been achieved with quite different physical systems, for example, with ions in ion traps, with superconducting contacts, or with impurities in diamond crystals (NV centers). To give an idea of the approach, let us consider a concrete example: the qubits in one of the first realizations of a quantum gate, achieved in 1995 at the National Institute of Standards and Technology (NIST) in Boulder. The qubit states were realized by the energy levels of a single beryllium ion in an ion trap.

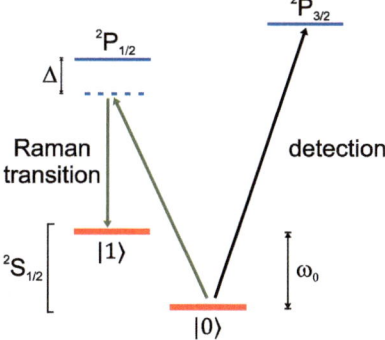

Figure 2.8: Energy levels used as qubit in an ion trap experiment (after [11]).

Fig. 2.8 shows the states used in the experiment. The horizontal lines represent the energy levels of the electrons in the beryllium ion. From a large number of levels, only those relevant to the experiment are shown. The energy of the states increases from bottom to top. The states are labeled in spectroscopic notation, which provides information about the angular momentum quantum numbers of the states involved. The two states used as qubit are labeled $|0\rangle$ and $|1\rangle$. This notation for quantum states is called Dirac notation. It is very useful for symbolic calculations and is explained in more detail in Section 3.8. The transition between the two states can be excited by microwave radiation at a frequency of 1.25 GHz (indicated by the double arrow labeled with ω_0). Such frequencies are typical for transitions in the hyperfine structure of atoms.

In the experiment considered, the transitions between qubit states were not excited directly with microwaves, but with lasers, because they are easier to handle experimentally. For this purpose, a higher energy state was used (labeled $^2P_{1/2}$). There were *two* lasers with wavelengths in the UV range that were directed at the beryllium ion. Their frequencies were not exactly the same but differed by the excitation frequency between $|0\rangle$ and $|1\rangle$ (green arrows). In this way, transitions between $|0\rangle$ and $|1\rangle$ can be induced by a "detour" via the $^2P_{1/2}$ state (this is what the term "Raman transition" means). In fact, the lasers are not precisely tuned to the excited state. They are detuned from resonance by the detuning frequency Δ (horizontal dashed line). As a result, the $^2P_{1/2}$ state is never truly excited, so the transition can effectively be thought of as taking place only between the two states $|0\rangle$ and $|1\rangle$. By choosing the laser pulses appropriately, one can induce transitions between $|0\rangle$ and $|1\rangle$ or create arbitrary superposition states to perform gate operations on the qubit.

Finally, another energy level comes into play: the transition between $|0\rangle$ and the $^2P_{3/2}$ state is used to measure the qubit. If the qubit is in the state $|0\rangle$, the laser frequency fits and the transition can be excited. The ion will then emit photons at the corresponding frequency.

Zeeman effect

As already mentioned, the term diagram in Fig. 2.6 on p. 27 shows an example of the *degeneracy* of energy levels with respect to the quantum numbers ℓ and m. Their energy does not depend on these quantum numbers. For most types of atoms and many other bound systems with discrete energy levels (e. g. NV centers), this changes when an ex-

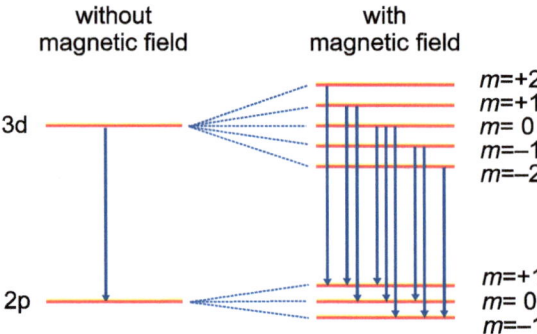

Figure 2.9: Zeeman effect: energy shift of states depending on magnetic field.

ternal magnetic field is applied. In a magnetic field, the energy of each level shifts by a certain value that depends on m. Thus, the degeneracy with respect to m is removed. This effect is known as the *Zeeman effect*. It is the basis for magnetic field sensors in quantum technologies. In the simplest case, the energy varies according to:

$$\Delta E_{\text{Zeeman}} \sim m \times B. \tag{2.4}$$

Depending on the structure of the couplings of angular momentum and spins within the atom and the strength of the applied magnetic field, however, the dependencies may be more complicated.

As a consequence of the Zeeman effect, spectral lines split into several components because more transitions with different frequencies are now possible. Figure 2.9 illustrates this line splitting using the example of a transition from a 3d to a 2p level (i. e., from $n = 3$, $\ell = 2$ to $n = 2$, $\ell = 1$). Both the higher and lower energy states are split into $2\ell + 1$ levels with different m.

ℹ️ In principle, transitions between all energy levels are possible. However, the probabilities differ so much that a distinction is made between "allowed" and "forbidden" transitions. Allowed transitions are those for which the *selection rules* $\Delta\ell = \pm 1$ and $\Delta m = 0, \pm 1$ are satisfied. These transitions are called *dipole transitions* and are shown in Fig. 2.9. The probability of forbidden transitions is orders of magnitude smaller. This is exploited in quantum technologies: To realize qubit states that are as stable as possible, one looks for long-lived states that are coupled to other states with lower energy only by forbidden transitions.

2.4 Josephson junctions and superconducting qubits

For successful use in quantum technologies, components must be low-noise and well-controlled. This is often the case in the quantum optical systems discussed above, but there are also appropriate systems in solid-state physics. Superconducting materials are well suited for this purpose: below a certain temperature, they conduct electricity without resistance, so one source of dissipation is eliminated from the outset.

In superconductivity, two electrons combine to form a bound state, a *Cooper pair*, which propagates through the solid without resistance. To break up a Cooper pair, normal superconductors require excitation in the millielectronvolt range. This energy corresponds to thermal excitation of about one Kelvin (hence the low temperatures required for superconductivity) or electromagnetic radiation at a frequency of 20 GHz (which is in the microwave range). In fact, superconducting qubits are cooled even lower, to about 20 mK, using liquid helium in cryostats and special cooling techniques.

Example: Show with an order-of-magnitude estimate that the energy transmitted by microwave photons with a frequency of 20 GHz corresponds to the thermal energy of an object with a temperature of 1 K.

Solution: According to the kinetic theory of gases, the average thermal energy of an object at temperature T is of the order of $k_B T$, where $k_B = 1.38 \times 10^{-23}$ J/K is the Boltzmann constant. According to Eq. (2.1), the photon energy at frequency f is given by $E = h \times f$. Equating the two formulas and solving for f gives:

$$f = \frac{k_B T}{h} = \frac{1.38 \times 10^{-23} \text{ J/K} \times 1 \text{ K}}{6.626 \times 10^{-34} \text{ Js}} = 20.8 \text{ GHz.} \tag{2.5}$$

An important element used in all superconducting qubit realizations is the *Josephson junction*. It consists of two superconductors separated by a thin insulating layer. Cooper pairs can tunnel through the insulator, i. e. they can cross it with a certain probability.

For a Josephson junction, the voltage across the insulating layer is proportional to the rate of change of the tunneling current. In a circuit, the Josephson junction thus behaves similarly to a coil. When it is connected in parallel with a capacitor, a resonant circuit is created, which forms the basis for superconducting qubits. The energy of the oscillations in the resonant circuit is quantized. Discrete energy levels are formed so that the energy differences between the levels are not all equal. The latter fact is crucial because to use them as qubit levels, it is necessary to be able to selectively excite them with microwaves. The Josephson junction, with its nonlinear properties, is responsible for the different spacings of energy levels. Linear components would result in equidistant energy levels that could not be addressed individually.

The exact design of superconducting qubits is determined by the lowest possible sensitivity to noise sources. The basic types are called *phase qubits*, *flux qubits*, and *charge qubits*. They differ in the shape of the potential curves and the energies that occur at the capacitance and inductance. There are several variants, of which especially the *transmon qubit* is used in the practical realization of quantum computers.

2.5 Light and photons

In the history of physics, conceptions of the nature of light have undergone multiple changes. Newton based his explanation of optical phenomena on a corpuscular theory (Opticks, 1704). However, since the interference experiments of Fresnel (1822), the wave nature of light was generally accepted. Its hallmark is the phenomenon of *interference*, where amplification or attenuation of intensity occurs when different parts of a light beam are superimposed coherently (i. e., with a fixed phase relationship). These effects are called constructive or destructive interference. In particular, the occurrence of destructive interference (in Fresnel's words, "*that in certain cases light, added to light, produces darkness*") seemed incompatible with any particle conception.

Subsequently, optics was fully embedded in the general theory of the electromagnetic field (Maxwell, 1865). The fundamental equations describing all phenomena in classical optics are *Maxwell's equations* – partial differential equations for the evolution of the electric field $\vec{E}$ and the magnetic field $\vec{B}$ in space and time.

In practical, applied optics, such as the construction of optical devices (lenses, microscopes, and other imaging elements), *geometrical optics* is still used. It describes the limiting case where the structures involved are so large that interference and diffraction phenomena do not play a role. Geometrical optics considers light rays that travel in straight lines in homogeneous media and only change direction in optical components like lenses or mirrors.

The established view of light as a wave was challenged at the beginning of the 20th century by Einstein's explanation of the *photoelectric effect*. Using a particle model, Einstein explained the fact that electrons can be kicked out of a metal surface by irradiation with light. In his explanation, he assumed that the energy transferred from the incident light to the metal surface is quantized, i. e. it comes in discrete portions. The energy is thus transferred by the light in the same way that it would be transferred by particles. The term *photons* was later coined for these particles of light. Einstein was able to quantitatively explain the photoelectric effect by assuming the relationship $E = h \times f$ between the energy and the frequency of a photon. He received the Nobel Prize in 1922 for his explanation of the photoelectric effect – not for the theory of relativity.

After Einstein's discovery, there was a debate about the "wave-particle dualism", the seemingly incompatible descriptions of light as a wave and as a particle. It was one of the central topics in the discussion about the interpretation of quantum physics. In Section 3.2, we will deal in detail with the resolution of the wave-particle dualism in terms of Born's probability interpretation.

Even if Einstein's explanation of the photoelectric effect has the merit of great conciseness, it is no longer considered evidence for the photon nature of light. Already in 1926, Wentzel showed that the photoelectric effect could also be explained by a semiclassical theory (quantized atoms interacting with a classical light wave).

Photons as excitations of the electromagnetic field

In modern physics, the term photon has a well-defined meaning that goes far beyond a naive conception of classical particles. The framework is provided by quantum electrodynamics – the quantum theory of the electromagnetic field. For a quantized description of light, one starts from the solutions of Maxwell's equations under the particular conditions of the experiment. These classical solutions of Maxwell's equations are commonly referred to as *modes*. They may describe laser pulses, standing waves in a resonator, modes in an optical fiber, or light propagating along the various paths in an interferometer. The modes form the basis of the photon concept.

In quantum electrodynamics, it turns out that the modes of the field can only be excited in a quantized manner. These quantized excitations are called photons. One mode can be occupied by one photon, by two photons, by three or more photons. A superposition of different numbers of photons is also possible – but not a non-integer occupation of the mode. If the energy content of a mode with frequency f is measured with a detector, only integer multiples of the Einstein value $E = h \times f$ are found.

Photons are quantized excitations of the electromagnetic field modes.

The quantum electrodynamical description of photons as modes of the electromagnetic field answers the frequently asked question of how to visualize a photon. The short answer: Think of it like the corresponding mode of the electromagnetic field, like the classical solution of Maxwell's equations. This means that a photon can have very different shapes: in an ultrashort laser pulse, it is concentrated in a very small space, while in a gravitational wave interferometer (where the associated modes are standing waves that extend throughout the interferometer), it can be stretched over kilometers.

In measurement with a detector, a photon is always found as a whole – all of its energy is transferred in its entirety during the detection process. When we describe the detection of quantum objects as a quantum mechanical measurement process on p. 45, we will summarize this fact in the rule of thumb *"a photon is wave-like in propagation and particle-like in detection"*.

Photon number statistics

Different light sources can be categorized according to the *photon number statistics* of the light they emit: the probability distribution $P(n)$ for the number n of photons that occupy a given mode. By repeated measurements with an ideal detector, the mean $\bar{n}$, the standard deviation Δn, and the shape of the distribution can be determined.

Fig. 2.10 compares three fundamentally different types of light according to their photon number statistics. The three distributions all have the same mean photon number $\bar{n} = 5$, but different shapes and standard deviations. The light brown distribution corresponds to thermal light emitted by a light source that radiates because it is heated (such as an incandescent lamp, a piece of charcoal, or the sun). The energy of thermal light is distributed according to *Planck's radiation law*. It is also called *classical light* because it is the kind of light that is emitted in nature by classical, macroscopic bodies. The light blue distribution shows the photon number distribution of a *coherent state*, which describes the light emitted by an idealized laser. It is characterized by a Poisson distribution with $\Delta n = \bar{n}$.

States of light that are narrower than the Poisson distribution (i. e. for which $\Delta n < \bar{n}$) are called *nonclassical states of light*. This type of light can only be emitted by controlled quantum systems such as single atoms, ions, or NV centers. An extreme variant of nonclassical light are the *Fock states*, which have a fixed photon number – in the case of

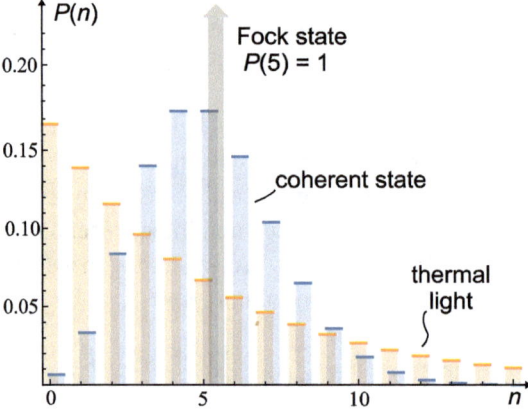

Figure 2.10: Photon number distribution for thermal light, a coherent state and a Fock state, each with the same mean photon number $n = 5$.

the gray distribution in Fig. 2.10, it is $n = 5$. Because of their nonclassical character, Fock states are difficult to generate experimentally.

Bunching and antibunching

Not only the distribution of photon numbers is different for classical and nonclassical light, but also the statistics of the arrival times at a detector. For any random process with independent events (such as raindrops hitting a tin roof), the statistical distribution of the events is a Poisson distribution. If you visualize such a random process by marking each point in time that a raindrop hits with a vertical bar, you get a distribution like the one in the middle row of Fig. 2.11. The same Poisson statistics is obtained for the detection of photons in a coherent state.

Classical light (e. g., from a thermal source) has a different distribution: the photons often appear "lumped"; the probability that two photons follow each other in a short time interval is larger than for Poisson statistics (Fig. 2.11(a)). This effect is called *bunching*; the distribution follows a *super-Poisson statistics*.

The opposite case, when two successive photons are on average more separated in time than in Poisson statistics, is called *antibunching* (Fig. 2.11(c)). The occurrence of antibunching (and the corresponding *sub-Poisson statistics*) is the defining characteristic of nonclassical light.

To determine the photon statistics experimentally, the correlation function $g^{(2)}(\tau)$ is measured. It describes the probability of detection for two photons that are separated by the time τ. For coherent states, $g^{(2)}(\tau = 0)$ is normalized to 1. In the case of nonclassical light, $g^{(2)}(0)$ is less than 1, and for classical light, it is greater than 1.

The occurrence of antibunching is easy to understand when the light source is a single atom: If the atom has just emitted a photon, then it will, with certainty, be in the ground state afterwards. It cannot emit another photon until the next excitation

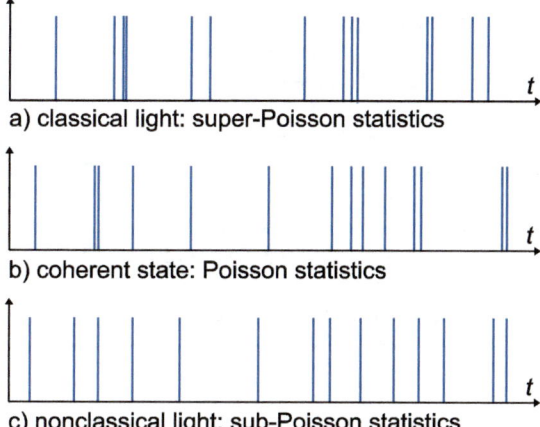

a) classical light: super-Poisson statistics

b) coherent state: Poisson statistics

c) nonclassical light: sub-Poisson statistics

Figure 2.11: Photon detection time statistics for classical light, coherent states, and nonclassical light. Classical light shows bunching, and nonclassical light shows antibunching.

occurs. Thus, the probability that two photons will be emitted from this system in the time between two excitations is zero. Therefore, the light from a single atom (or NV center) is characterized by $g^{(2)}(0) = 0$.

Single-photon states and attenuated light

Photon statistics also explains why it is not possible to produce nonclassical light by attenuating thermal or laser light. A gray filter used for attenuation (which works no differently than sunglasses) randomly absorbs incoming photons; only some of them are transmitted. The fundamental problem with this approach is that by randomly removing events, you can attenuate the intensity, but you cannot change the photon statistics. The sub-Poisson statistics that characterizes nonclassical or single photon states cannot be achieved in this way. Attenuated thermal light still shows bunching instead of antibunching; attenuated laser light still follows Poisson statistics. In a pulse produced in this way, containing, on average, a single photon, there is still a probability (given by the corresponding statistics) of finding two photons, three photons, or no photon at all.

In practice, the heralded single-photon method described on p. 15 is usually used for the controlled generation of single-photon states. Here, pairs of photons are generated in a nonlinear crystal. One of them is used to "announce" the second photon. For practical reasons, however, attenuated laser sources are often used in quantum communication. In doing so, one has to be aware that these are not controlled single-photon states.

Detection of single photons

For the detection of single photons, avalanche photodiodes (APDs) are usually used (Fig. 1.11). They work like a photomultiplier, but on a semiconductor basis. An APD has a layered structure in which one zone is under high voltage. An incident photon creates an electron-hole pair that is accelerated by the voltage. Similar to the formation of an avalanche, many additional ionizations are triggered by impact ionization so that, in the end, an electrical signal can be measured. APDs can have quantum efficiencies close to 100 % (which would mean they reliably detect every incoming photon). However, the quantum efficiency is wavelength-dependent and can be further reduced by filters that are necessary in some experiments.

However, even without an incident photon, an electron-hole pair can be created by thermal excitation. These unwanted events result in the *dark count rate*, the noise floor of APDs. Heralded single photons are used to suppress this noise and to distinguish the random events from actual photon detections.

Single-photon detectors can have other errors that bias the detection statistics. For example, they have a certain *dead time*: After an excitation has occurred in the APD, no further detection is possible for some time because the excitation has to be "cleared" first. Other errors are the triggering of two pulses by one single photon (afterpulsing) and a time delay between the arrival of the photon and the output signal.

3 Using quantum physics

3.1 The double-slit experiment

Since the legendary *Feynman Lectures* of 1963, the *double-slit experiment* has been considered the key experiment for understanding quantum physics. In this experiment, many of the "oddities" of quantum physics appear in their purest form in a simple experimental setup. It turns out that the experimental results cannot be reconciled with our familiar ideas from classical physics. Our imagination is not capable of providing a consistent, intuitive description of what is happening in this experiment.

Feynman himself describes the double-slit experiment as „[...] a phenomenon which is impossible, absolutely impossible, to explain in any classical way, and which has in it the heart of quantum mechanics. In reality, it contains the only mystery" [12]. With this statement, however, he does not want to say that quantum physics is fundamentally mysterious, without any rules, and incomprehensible. The published Feynman Lectures are based on transcribed audio recordings of the original lectures, and the following crucial passage was not transcribed ([12] at 8:17 min):

> All the peculiarities of quantum mechanics are always the same thing – so I have selected a particular experimental situation to describe which contains the peculiarities of quantum mechanics in a cute form, and any other place where you have peculiarities is just another example of the same thing – so we are right at the heart of the subject.

Thus, according to Feynman, quantum physics has a limited number of oddities. They can be demonstrated using the example of the double slit and then transferred to other cases. This is the benefit of the double-slit experiment: We can learn from it the rules that help us in other, more complex situations.

Double-slit experiment with classical light

The basic setup of a double slit experiment is shown in Fig. 3.1. Light from a point source S passes through two narrow slits and then falls onto a screen. A characteristic *interference pattern* with light and dark fringes appears. Its intensity profile is shown on the right side of the figure. Compared to a uniform distribution, the intensity is higher at some points (interference maxima) and lower at others (interference minima). The appearance of interference minima is especially remarkable because here the relation "bright+bright=dark" applies. At these points, the intensity is lower with two slits open than with only one slit open.

The double-slit experiment was first described by Thomas Young in 1803 to settle the dispute between the wave and particle theories of light. Its result provided evidence for the wave theory because the appearance of interference is the typical characteristic of superimposed waves (as indicated in Fig. 3.1 behind the two slits). Interference can hardly be explained with a classical particle conception.

The peculiarities mentioned by Feynman do not yet appear in this variant of the double-slit experiment, which is performed with "bright" light (from the sun, a lamp, or a laser). Its results can be easily explained using classical optics.

https://doi.org/10.1515/9783110717457-003

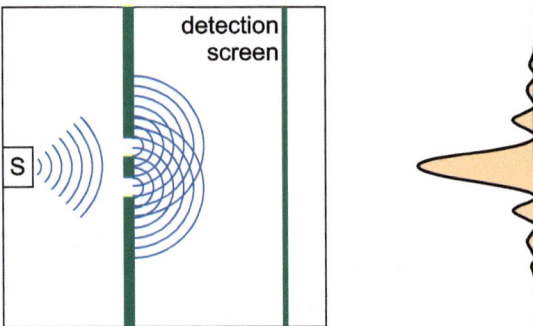

Figure 3.1: Basic setup of the double-slit experiment.

The peculiarities occur only when the double-slit experiment is performed with single quantum objects – for example, single photons, electrons, whole atoms, or molecules. The exact nature of the quantum objects used does not matter because, as Feynman puts it, the peculiarities are always the same.

ℹ Double-slit experiment with single atoms

In 1991, a double-slit experiment with whole atoms was realized for the first time [13]. Its setup corresponded exactly to the prototypical scheme shown in Fig. 3.1. Single helium atoms were made to interfere by passing through a double-slit. A thin gold foil with two 1 μm wide slits 8 μm apart served as the double slit arrangement (bottom right of Fig. 3.2).

In order for the helium atoms to propagate freely, the experiment had to take place in a vacuum. Before passing the double slit, the atoms were excited by electron collisions. Behind the double slit, they hit a gold foil that served as a detection screen. By releasing their excitation energy, the helium atoms caused the emission of electrons. These electrons were registered in a spatially resolved manner using electron optics. Figure 3.2 shows the pattern of the detected helium atoms at different times during the 42 hours of the experiment.

When a helium atom is detected, its energy is released at a spatially well-defined "spot" on the gold foil. At first, the spots are apparently randomly distributed (first image in Fig. 3.2). As the number of detected atoms increases, a pattern gradually forms from the traces of the individually detected atoms. It has the same structure as the interference pattern in the double-slit experiment with light.

3.2 The basic rules of quantum physics

The result of the double-slit experiment with helium atoms cannot be explained by our classical, intuitive notions of waves or particles. We will not discuss in detail where the classical conceptions fail. Instead, in the spirit of Feynman, we will use the experiment to read off the limited set of rules that quantum physics seems to follow. With these rules, we will be able to discuss a variety of applications in quantum technologies.

It turns out that a set of four qualitative rules is sufficient to describe most, if not all, of the peculiar features of quantum physics. They are called the *basic rules of quantum physics* [14] or *reasoning tools* [15]. They are qualitative statements that are supposed to contain the basic features of quantum physics and provide a kind of "qualitative mini-

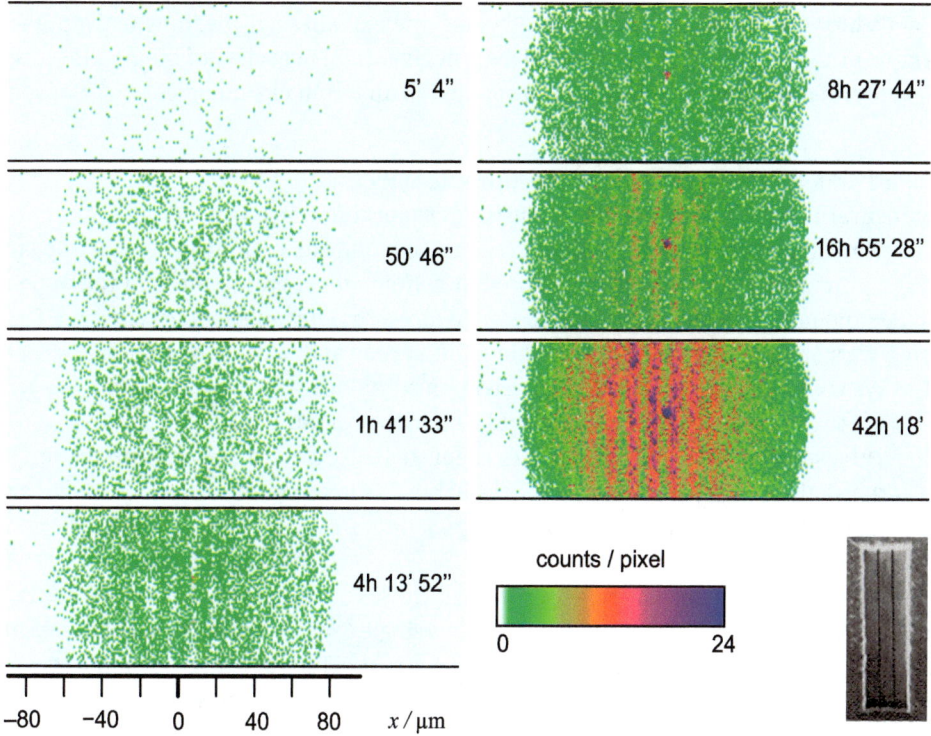

5' 4"

8h 27' 44"

50' 46"

16h 55' 28"

1h 41' 33"

42h 18'

4h 13' 52"

counts / pixel

0   24

−80   −40   0   40   80 x / μm

Figure 3.2: Double-slit experiment with helium atoms (data: Christian Kurtsiefer).

axiomatics". They aim to serve as an aid in qualitative discussions of quantum physics and, in particular, to build up an "intuition" for quantum phenomena. They read as follows:

Rule 1: Indeterminism and statistical predictability
Single events are not predictable, they are random. Only statistical predictions (for many repetitions) are possible in quantum physics.

Rule 2: Interference of single quantum objects
Interference occurs if there are two or more "paths" leading to the same experimental result. Even if these alternatives are mutually exclusive in classical physics, none of them will be "realized" in a classical sense. Quantum states showing interference are called *superposition states*.

Rule 3: Definite measurement results
Even if quantum objects in a superposition state need not have a fixed value of the measured quantity, one always finds a definite result upon measurement.

Rule 4: Complementarity
Exemplary formulations are: "Which-path information and interference pattern are mutually exclusive" or "Quantum objects cannot be prepared for position and momentum simultaneously".

Let us now discuss the basic rules one by one and explain their meaning using the example of the double-slit experiment. This will give us an overview of the non-intuitive features of quantum physics that are exploited in quantum technologies.

Rule 1: Indeterminism and statistical predictability

Statistical behavior is perhaps the most important difference between quantum and classical physics. In the double-slit experiment, it manifests itself in a way that appears quite unremarkable at first glance. The first helium atoms hitting the gold foil appear to be randomly distributed (fig. 3.2 top left) before a recognizable pattern emerges. Suppose we stop the experiment at a certain point in time and make two predictions:

1. We try to predict the location where the next helium atom will be found. We will not succeed. Such a prediction would be a matter of pure luck.
2. Things look different if we consider the next 1000 detected atoms: A prediction of the statistical distribution succeeds reliably. We can reproducibly indicate where many atoms are detected and where few.

The statement made in basic rule 1 directly addresses the difference between these two cases: We have moved from a statement about a single event to a statement about repeating the same experiment many times, i. e., to a statistical statement. Only in the second case a prediction is possible. This prediction is reproducible: Each time the experiment is repeated with many helium atoms under the same conditions, one can predict – except for statistical fluctuations – the distribution of the atoms. In short, the laws of quantum physics are *probabilistic*. Deterministic predictions about individual events are generally not possible.

 Quantum physics and determinism

Comparison with classical physics can further clarify the probabilistic character of the laws of quantum physics. Classical mechanics is deterministic. If you shoot an arrow at the right speed and angle, you can be sure it will hit the target. Two arrows with identical initial conditions will follow identical paths (Robin Hood's magic trick).

In quantum physics, the process of setting up well-defined initial conditions is called *preparation*. And it turns out: In the double-slit experiment, aiming does not work. It is not only impractical but, in principle, impossible to prepare the atoms in such a way that one particular atom will land on the gold foil at a predetermined position. This is not just a matter of insufficient control over the initial conditions. Quantum physics is *indeterministic* in principle. For a single event, identical initial conditions do not lead to identical results. Two atoms prepared in the same way will usually end up in different places on the screen. Only probabilistic statements can be made about the location where they are detected.

Although only statistical predictions are possible, the laws of quantum physics are strictly valid laws of nature. The difference to classical mechanics is that in quantum physics, the laws are probabilistic. They do not determine the outcome of single events (e. g., the location of a single atom) but describe a whole series of experiments: Many repetitions of the same experiment result in a distribution that is reproducible (except for statistical fluctuations).

One might object: Even in classical physics, there are processes whose outcome is apparently determined by chance. An example is a roll of the dice. However, if one knew the experimental conditions, includ-

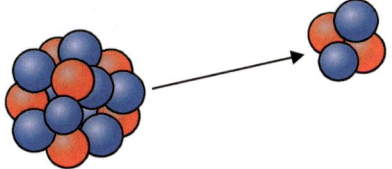

Figure 3.3: The radioactive decay of an atomic nucleus as an example for the indeterminism of quantum physics.

ing the initial conditions (the air movements, the unevenness of the surface, etc.), one could, in principle, determine the outcome with Newton's laws. Thus, the result of the dice roll is basically determined from a Newtonian point of view; the chance results from our ignorance of the exact initial conditions.

If, on the other hand, we cannot predict the detection location of a helium atom in the double-slit experiment, it is not a matter of subjective ignorance of the initial conditions, but a fundamental limitation. According to quantum mechanics, there is no additional parameter that determines in advance where a particular atom will be found on the screen. No better control of the atoms is possible.

Bell's theorem, which we will discuss in section 3.13, shows that this is not an inadequacy of quantum mechanics that could be overcome by a better and more complete theory. Bell's theorem shows that theories with "hidden parameters", which predetermine the location of the detection, must have explicitly non-local properties.

A final remark on the statistical character of quantum mechanical laws: Of course, it is not impossible that the probabilities 0 or 1 occur. In many experiments, one will even actively try to obtain probabilities close to 1 for a given event. In these cases, one can quite confidently predict the occurrence or non-occurrence of individual events.

A particularly striking example of the statistical nature of quantum laws is the decay of radioactive nuclei. In alpha decay, for example, an atomic nucleus emits an α particle (Fig. 3.3). It is impossible to predict whether a given individual radioactive nucleus will decay within the next hour. There is no parameter that determines the occurrence of this event in advance. Nuclear decay is an example of true randomness in quantum physics.

However, it is possible to make a statistical prediction: We are able to tell what fraction of a very large number of nuclei will decay within the next hour. We can formulate a reproducible statistical law, the exponential decay law, with the half-life characteristic for the nuclear species.

In quantum technologies, the probabilistic and indeterministic character of quantum processes is exploited directly, for example, in quantum random number generators or in secure key distribution in quantum communication. On the other hand, quantum randomness complicates the design of algorithms for quantum computers.

Rule 2: Interference of single quantum objects

We have already seen that although the individual helium atoms are always detected at only one location, their points of impact form a double-slit interference pattern after many repetitions. Remarkably, the interference pattern occurs even when there is only a

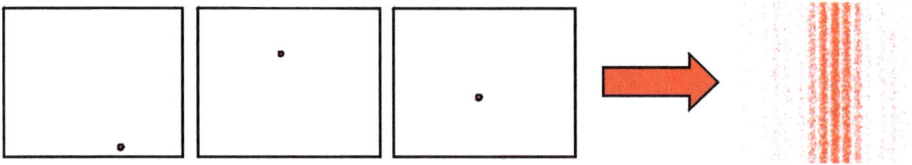

Figure 3.4: The "superposition" of single events leads to the interference pattern.

single quantum object in the arrangement at a given time so that any mutual interaction is excluded. Paul Dirac, one of the founders of quantum mechanics, put it this way: Each quantum object interferes only with itself.

There was also no mutual influence of the atoms in the helium double-slit experiment of Fig. 3.2. The long period of 42 hours meant an average time between two detections of about three seconds. Despite the large number of atoms detected in total, each atom was registered individually, without any of the other detected atoms being present in the apparatus on average.

ⅈ **Each quantum object interferes only with itself**

Dirac's statement can be illustrated very vividly: An organization sends identical blueprints for a double-slit interference experiment to many laboratories worldwide. The apparatus is built according to the specifications in each of the laboratories. A single experiment is performed with each of the identical apparatuses – with only a single atom in each laboratory. The location of this atom is drawn on an overhead transparency (Fig. 3.4 left). At a conference, all the participants come together, each with a slide on which there is a dot. They place their slides on top of each other on the projector. The result is the double-slit interference pattern (Fig. 3.4 right).

Rule 2 gives a criterion for the occurrence of interference: Whenever there are several "possibilities" for the occurrence of a certain experimental result (more generally: classically conceivable alternatives for the realization of the same experimental result), interference can be detected. In the double-slit experiment, the experimental result is the detection of the atom at a certain location X on the gold foil. There are two classically conceivable alternatives for the realization of this result: the atom can get there through the left slit (alternative 1) or through the right slit (alternative 2). The experimental result, the detection at position X, is the same in both cases.

Quantum mechanically, the state of the atoms in the plane of the slits is described as a *superposition state* of the two classical alternatives. This is symbolically expressed by the following notation:

$$|\psi\rangle = |\text{State for alternative 1}\rangle + |\text{State for alternative 2}\rangle . \qquad (3.1)$$

Such a state is the prototype of a "qubit state", where neither of the two classical alternatives is actually realized. Superposition states are very common in quantum technologies; in fact, they are their hallmark. Usually, it is not the "left slit" and "right slit"

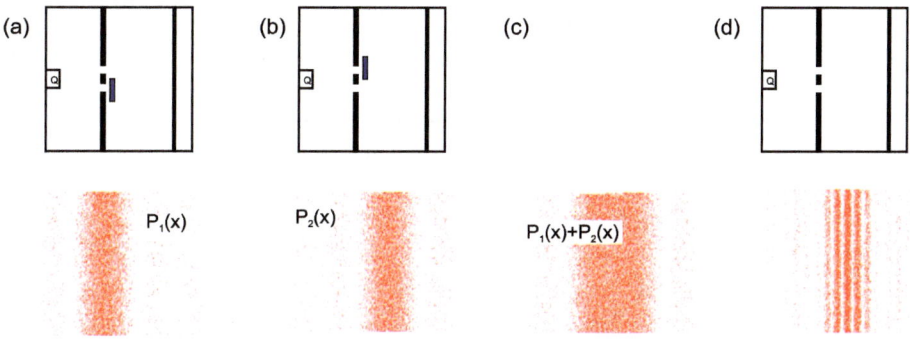

Figure 3.5: "Rearranging" the atoms (computer simulation).

alternatives that are superposed, but the internal states of quantum objects, commonly denoted $|0\rangle$ and $|1\rangle$. However, the superposition state considered here in the double-slit experiment has the advantage that it can be used to justify very clearly the often-heard but rarely explained statement that a qubit in a superposition state of $|0\rangle$ and $|1\rangle$ is neither in one state nor in the other, but "in both states at the same time" – and to answer the question what this actually means.

Superposition states and the "non-attribution of properties"

Within the double-slit apparatus, the atoms are in a "qubit" state: a superposition state of "went through the right slit" and "went through the left slit". Neither of the two classical alternatives is actually realized in this case. This is revealed – as we will show in a moment – in the occurrence of interference.

Let us assume the opposite, that one of the alternatives, "left slit" or "right slit", was actually realized. Each helium atom could then be assigned to one of the two slits through which it passed – we just would not know which. Such a state could be called an "either-or" state [16].

The following argument shows that this "either-or" conception does not describe our experiment: If the assumption were true, then the pattern seen on the screen should remain unchanged if we "rearrange" the atoms. To do this, we first let pass all those atoms which go through the left slit and only afterwards those which go through the right slit. This rearrangement of the atoms can be done by first closing the right slit, then reopening it and closing the left slit.

The resulting pattern on the screen is shown in Fig. 3.5. The atoms that have passed through the left slit produce the distribution $P_1(x)$ on the gold foil; those passed through the right slit produce the distribution $P_2(x)$. Both distributions taken together give the pattern $P_1(x) + P_2(x)$ shown in Fig. 3.5(c). However, contrary to our assumption, this distribution does not match the one obtained in an experiment with two open slits (Fig. 3.5(d)).

Thus, there is a mistake in the chain of argumentation. We run into difficulties if we imagine the atoms inside the apparatus as objects with a well-defined position or trajectory. This argument explains the statement in rule 2 that neither alternative is realized in the classical sense. Superposition states can be distinguished from classical "either-or" states by the occurrence of interference. The argument also invalidates the "either-or" description of quantum objects in superposition states.

Superposition states and interference are central to quantum technologies: the advantage of quantum computers compared to classical computers is essentially made possible by using qubits in superposition states and the occurrence of interference.

As an aside, it should be added that "either-or" states also exist in quantum mechanics. They are called *mixed states* – not surprising after the above description of how they are created by "mixing" the classical alternatives. Mixed states differ from superposition states by the absence of interference (cf. section 3.11).

Rule 3: Definite measurement results

As we have seen, we cannot assign specific properties to quantum objects in superposition states – for example, a specific slit to the atoms in the double-slit experiment. More generally, we cannot assign them a specific position in certain experimental situations.

One might object that, after all, the position of an atom can be measured. What is the result of a position measurement when the measured atom is in a delocalized state, as in the double-slit experiment? We already know the answer from the experimental result in Fig. 3.2. Here, a position measurement is made by the detection on the gold foil, where the helium atoms release their excitation energy and trigger an electron. The experiment shows: In a position measurement, the atom is found at a definite position – even if it was previously in a state where it was delocalized at least over the distance of the two slits (8 μm in the experiment).

This result is an example of a general feature of measurements in quantum physics: In a measurement, one always finds a definite value of the measured quantity. This is true even if the quantum object is in a superposition state before the measurement, in which the property in question cannot be assigned to the quantum object at all.

In this respect, a quantum measurement differs from measurements in classical physics: In classical physics, the measurement reveals a pre-existing value of the quantity being measured. In contrast, a quantum measurement has a more active character. The measured system is "forced" to "decide" for one of the possible measurement results.

 Qubit measurements

Let us look at the same topic in the context of qubits. We consider a qubit that is in a superposition state of $|0\rangle$ and $|1\rangle$. In quantum computing, $|0\rangle$ and $|1\rangle$ are called the *computational basis*. Generally, the state of the qubit can be written:

$$|\psi\rangle = \alpha\,|0\rangle + \beta\,|1\rangle . \tag{3.2}$$

Compared with Eq. (3.1), the coefficients α and β have been introduced. They are numbers (generally complex) that express that in $|\psi\rangle$, the two components $|0\rangle$ and $|1\rangle$ do not appear in equal parts, but with different weights. When a quantum algorithm is executed in a quantum computer, the qubits are put into such superposition states; usually, the states are even more complex superposition states involving multiple qubits. This is the basis of *quantum parallelism*.

At the end of the calculation, the qubit has to be read out. For this purpose, a measurement is performed. The measured quantity is the qubit value on the computational basis. According to rule 3, this measurement gives a definite result (0 or 1), even if before the measurement, when the qubit was in the superposition state (3.2), one could not assign a fixed value to it. We have already seen the experimental evidence for the definiteness of the measurement result: in the quantum jump experiment in Fig. 2.5, we can clearly see the jumps between the measurement results 0 and 1.

Using the formalism of quantum mechanics, which we will discuss in the following section, we can go one step further and specify probabilities for both measurement results. According to rule 1, the laws of quantum physics are statistical in nature. For the superposition state (3.2), we can calculate the probabilities for the two possible outcomes 0 and 1: they are $|\alpha|^2$ for the value 0 and $|\beta|^2$ for the value 1. So, if $|\alpha|^2 = 0.8$ and $|\beta|^2 = 0.2$, then in 1000 measurements, the value 0 is found about 800 times and the value 1 about 200 times. This interpretation of the coefficients α and β of Eq. (3.2) is a special case of Born's probability rule, which we will discuss on p. 56.

Wave-particle duality

In the early days of quantum mechanics, "wave-particle duality" was intensely debated. It was the first example of the fact that some terms of our language (such as "wave" or "particle"), which are derived from our experience in the macroscopic world, are not well suited to describe quantum phenomena. It should be noted that these problems have only ever affected our way of speaking about quantum phenomena. In a mathematical formalism, there have never been any difficulties correctly describing the experiments.

In the course of time, conceptually clearer ways of speaking about quantum physics have been developed. The basic rules discussed above are one example. They can be used to explain the "duality" between wave and particle, which seems so puzzling at first sight ("quantum objects sometimes behave like waves and sometimes like particles"). With rules 2 and 3, one can unambiguously predict for any experiment whether one will find "wave behavior" or "particle behavior" or both.

According to rule 2, the propagation of quantum objects in space and time is wave-like. Experimentally, wave behavior manifests itself in the occurrence of interference. This happens if there are different "paths" (like the slits in the double-slit experiment or arms of an interferometer) to realize the same experimental result. Of course, there are many experiments in which the apparatus does not contain different paths. Then, there will be no interference, even though the propagation of quantum objects is, in principle, governed by wave-like laws.

Particle-like behavior will only occur during a measurement. If we make a position measurement (e. g., with the gold foil in the helium double-slit experiment), then according to rule 3 we will find a definite value for the position. The quantum object is detected as localized, i. e., particle-like.

As a rule of thumb, we can say that quantum objects propagate like waves and are detected like particles. The wave-like propagation is completely deterministic. It is described by the Schrödinger equation, the fundamental dynamical equation of quantum mechanics, which resembles a wave equation (since it does not play a major role in quantum technologies, it is only mentioned here in passing). On the other hand, particle-like behavior and the indeterministic, probabilistic character of quantum phenomena manifest themselves only in measurements. They are described by Born's probability rule (which is related to rule 3).

Something similar can be formulated for qubits: The qubits in a quantum computer behave "analog" during the computation (continuous superposition of $|0\rangle$ and $|1\rangle$ according to Eq. (3.2), with the possibility of interference). Only the measurement at the end of the calculation gives a "digital" result: one of the two possible values 0 or 1 is found, each with a certain probability.

Rule 4: Complementarity

The term "complementarity" was coined by Niels Bohr. It encompasses a whole bundle of relations between two different variables, all of the form, "you can have one or the other, but not both at the same time." The best-known example is the *Heisenberg indeterminacy relation* for position and momentum. Similar relations, which are more relevant in quantum technologies, also apply, for example, to the polarization components of light (cf. Section 3.7).

At this point, we will stay with the double-slit experiment and focus on the complementarity between "which-path information" and "ability to interfere", which will lead us to the important topic of *environment-induced decoherence.*

We have already seen that there is no interference in the double-slit experiment if it is possible to assign a particular slit to the atoms – that is, if you have "which-path information". This statement can be made even more precise. To prevent interference, it is sufficient for the quantum objects to leave a trace somewhere in the environment, from which one could, in principle, infer which of the classical alternatives has been realized.

Such a trace could be, for example, a photon emitted by the helium atoms at some point along their trajectory. If this is the case, the experimental result mentioned in rule 2 is no longer "the detection of a helium atom at point X", but "the detection of a helium atom at point X and the emission of a photon with origin Y". If the slit through which the helium atom must have passed can be deduced from the origin of the photon Y, then there are no longer two alternatives for the realization of the experimental result: no interference occurs. The emitted photon does not even have to be registered by a detector. It is sufficient that it carries information about its origin to the environment.

The description of this experiment admittedly sounds a bit fantastic. However, it was actually performed by Pfau et al. [17] as a variant of the helium double-slit experiment a short time after the original experiment.

---

i **Which-path information and interference**

Unlike the helium experiment described above, the experiment was not performed with a physical double slit, but the atoms were diffracted by a standing light wave (Fig. 3.6). Due to the different light intensities in the nodes and antinodes, the standing wave acts like a diffraction grating. In the first step, the experiment was performed without photon emission. The laser that created the light grating was tuned to an off-resonant frequency that did not induce any transitions in the helium atoms. Accordingly, the atoms left the standing wave in the ground state so that they did not emit a photon on their way to the detector. This resulted in a clearly visible interference pattern, as shown by the data points drawn as open circles in Fig. 3.6.

In the second step, the laser frequency was tuned to a transition frequency of the helium atom. The result of this experiment is shown in Fig. 3.6 by the filled data points. We recognize an interference pattern that is reduced in contrast. This result can be explained as follows: By tuning the laser frequency to a resonance, some of the helium atoms are excited. In fact, the laser not only excites the atoms, but with the same probability, induces a return to the ground state. As a result, half of the helium atoms are in the excited state, and half are in the ground state.

The atoms in the ground state still show interference. However, the atoms leaving the standing light wave in the excited state will emit a photon on their way to the detector, from which their path could, in principle, be inferred. Thus, these atoms do not contribute to the interference pattern. The attenuated interference pattern shown in Fig. 3.6 contains both contributions: interfering atoms that have not emitted a photon and non-interfering atoms that have carried information about their path to the environment by emitting a photon.

The experiment clearly illustrates the complementarity of which-path information and interference pattern: one can get one or the other in an experiment, but not both at the

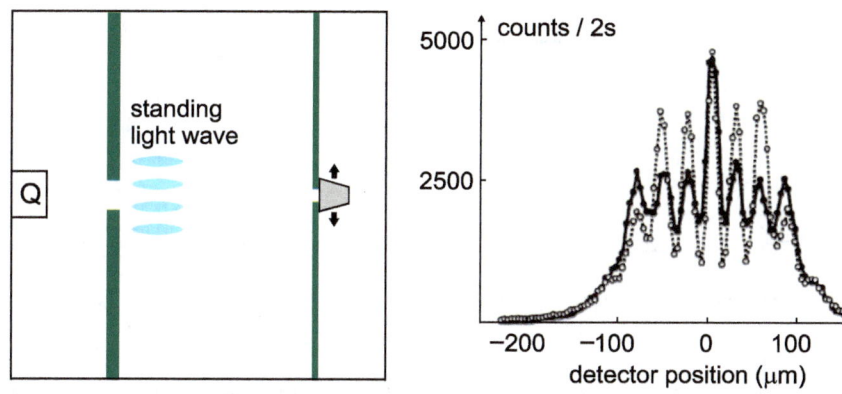

Figure 3.6: Interference of helium atoms without (open circles, dashed line) and with (filled circles, solid line) which-path information [17].

same time. This is true even if only partial path information is obtained: the contrast of the interference pattern then decreases accordingly [18].

Environment-induced decoherence

The experiment of Pfau et al. leads directly to the topic of *environment-induced decoherence*, which is eminently important for quantum technologies. The term denotes the loss of coherence (i. e., the loss of the ability to interfere) due to uncontrolled interactions with the environment. We have seen that the emission of a single photon, which allows distinguishing between the two alternative paths, was sufficient to prevent interference. The photon does not have to be detected for this purpose. It is sufficient for the suppression of interference that "which-path information" is transferred to the environment.

Virtually all applications in quantum technologies are based on using superposition states and the associated interference effects. Superposition is the very essence of quantum mechanics and, after all, the basis of the quantum advantage. If interference is suppressed by decoherence, the quantum advantage disappears. The quantum systems become "effectively classical"; they behave as if they were in classical "either-or states".

Therefore, it is essential to avoid environmental decoherence in quantum technologies. This is difficult, however, because every quantum system interacts with its environment to some extent: by the ever-present thermal radiation, by spontaneous emission, by scattering of light and gas molecules, or by interaction with lattice vibrations in solids. Measures against these uncontrollable interactions are often elaborate: for example, low temperatures to avoid thermal radiation or vacuum to avoid collisions with gas molecules. The physical qubit realizations used today were chosen for their insensitivity to decoherence. However, the shielding must not go so far as to prevent the manipula-

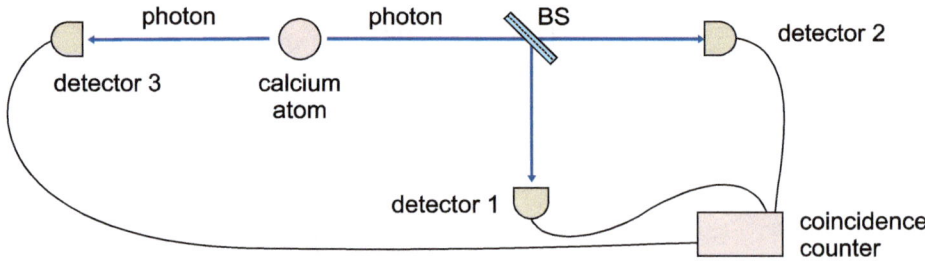

Figure 3.7: Anticoincidence experiment of Grangier, Roger, and Aspect [19].

tion of the quantum systems. Their states must still remain under controlled external influence. For example, NV centers or ions in traps are insensitive to thermal radiation in the infrared range (they do not have excitable transitions there). They can, however, be manipulated in a controlled manner by means of laser or microwave radiation.

3.3 Anticoincidence experiments with single photons

Besides the double-slit experiment, there is a second experiment that leads to the "heart of quantum mechanics". It was performed in 1986 by Grangier, Roger, and Aspect, and it is an early example of the use of heralded single photons (cf. p. 15). The experiment has two parts: In the first part, single photons hit a semitransparent mirror (also called beam splitter, BS). They are then registered by two detectors in an anticoincidence experiment (Fig. 3.7). In the second part, a second beam splitter is added to the experimental setup to create an interferometer. We will see that in this way, it is possible to obtain both characteristic particle and characteristic wave behavior in the same experiment by making only small changes in the experimental setup.

As the basic setup is so simple, this experiment is well suited to illustrate the peculiarities of quantum physics. We will first explain the results of the experiment qualitatively with the basic rules and then use it to give a first introduction to the quantum mechanical formalism.

ℹ Experiment by Grangier, Roger, and Aspect (the first part)

The setup of the experiment is shown in Fig. 3.7. The main difficulty at the time was creating single photons in a controlled way. To do this, the atoms of a calcium atomic beam were excited by lasers. They returned to the ground state by emitting *two* photons. One of the emitted photons served as a "trigger" photon. It was detected by detector 3 and announced the arrival of the second photon. Only when detector 3 had been excited, detectors 1 and 2 were enabled for a short time. In this way, it was possible to work with single photons in a controlled way.

The same principle is still used today to generate single photons, but with a different source. Instead of the cumbersome atomic beam, which requires a vacuum, the photon pairs are generated with much less effort in BBO crystals (cf. p. 15).

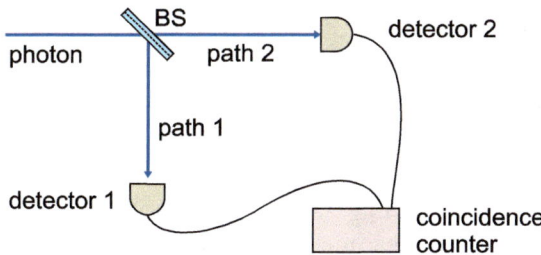

Figure 3.8: Single photons at a beam splitter.

Anticoincidence of single photons at a beam splitter

The basic idea of the first part of the experiment is simple (Fig. 3.8): A single photon hits the beam splitter BS. It has two possibilities there. It can be transmitted and then be registered by detector 1, or it can be reflected and then be registered by detector 2. Since it is a single photon, it should always be found in only one of the two detectors; the two detectors should never respond simultaneously. Thus, a coincidence counter connecting the two detectors would have to find perfect anticoincidence.

This result sounds plausible and not very surprising at first glance, and yet the observation of anticoincidence is genuine evidence for the photon nature of light. In fact, a classical wave would be split uniformly at the beam splitter. There would be a certain probability that both detectors would be excited simultaneously. Random fluctuations in light intensity would reach both detectors at the same time and would even increase the probability of a simultaneous excitation.

The experiment clearly confirmed the prediction of quantum physics: The coincidence rates of both detectors 1 and 2 were consistent with perfect anticoincidence.

Qualitative explanation with the basic rules

To demonstrate the explanatory power of the basic rules, we use them to explain the results of the experiment qualitatively. Initially, the beam splitter brings the photon into a superposition state of path 1 and path 2. In this superposition, the photon cannot be assigned any path. It is non-localized and distributed over both paths (cf. the reasoning in the box on p. 43).

According to the rule of thumb "wave-like propagation and particle-like detection", the photon can be thought of as a pulse propagating at the speed of light and split into two parts at the beam splitter – except that it is not a classical electromagnetic wave, but something more abstract, best described by the term *probability wave*. Both parts of the pulse are *jointly* occupied by a single photon (rule 2). This becomes apparent during the measurement.

Detectors 1 and 2 perform a position or "path" measurement of the photon. According to rule 3 (definite measurement results), the position measurement has a definite result. Exactly one of the two detectors responds, never both. The energy of the photon is transferred in its entirety to one of the two detectors. Thus, a single photon, when measured, is always found as a whole at a definite position, never at two positions at the same time. This corresponds to the anticorrelation found in the experiment.

It is unpredictable at which of the two detectors a particular photon is found. This follows from rule 1 (indeterminism and statistical predictability). According to quantum physics, there is no physical variable that determines in advance at which of the two detectors the photon energy will be registered. It is a matter of chance. Here, the indeterminism of quantum physics becomes apparent. Nevertheless, the process is not complete without regularities because a statistical prediction is possible. It is very simple in this case: With a 50/50 beam splitter about half of the photons are found in detector 1 and half in detector 2.

---

i **Quantum random number generators**

One of the first commercial applications of quantum technologies involved randomness. *Quantum random number generators* can be used to generate random numbers that are truly random. That does not sound very impressive. After all, routines for generating random numbers are part of every compiler in classical computing. However, it is important to remember that a classical computer is a deterministic machine. It *cannot*, in principle, produce true random numbers because the concept of randomness is completely at odds with its design principles.

Classical computers use deterministic routines that, when applied repeatedly, produce sequences of numbers that cannot be distinguished from true random sequences in practical use. Such routines are called pseudorandom number generators. Since the number sequences are calculated by a deterministic algorithm, they are not truly random. The sequence can be predicted if the starting value is known (often, the current time is used) and reconstructed accordingly later.

Even more serious is the fact that the number sequences returned by the algorithms are generally periodic. The length of the period depends on the starting values. This limitation is so serious that the authors of the standard work *Numerical Recipes* [20] formulated in a drastic way: "If all scientific papers whose results are in doubt because of bad [random number routines] were to disappear from library shelves, there would be a gap on each shelf about as big as your fist."

This is not the case with random number sequences produced by quantum randomness. If "transmitted" is encoded as 0 and "reflected" as 1 at a 50/50 beamsplitter, the result is a truly random sequence of bits. This kind of inherent quantum randomness is used in commercial quantum random number generators, which have been on the market since the early 2000s. They are used, for example, in cryptographic applications where "even small statistical errors in the keys have a negative impact on security" [21]. Online casinos are also obviously interested in ensuring that their random numbers are unpredictable and not pseudorandom. Many of them use quantum random number generators.

Second part of the experiment: Mach-Zehnder interferometer

While in the first part, mainly the particle aspects of light were emphasized, the experiment was modified in the second part to show mainly the wave properties. An interference experiment was carried out with heralded single photons from the same source as before. Although there was only one photon in the apparatus at any given time, interference could be observed – a typical wave phenomenon.

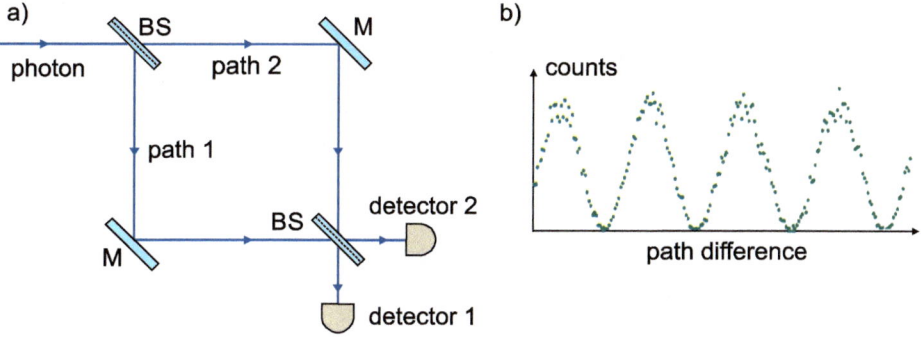

Figure 3.9: Extension of the experiment to the Mach-Zehnder interferometer (data: [19]).

Experiment by Grangier, Roger, and Aspect (The second part: Interference)　　　　　　**i**

In the second part, the experiment was modified by adding another beam splitter and two mirrors (M) to form an interferometer setup (Fig. 3.9(a)). In the particular type of interferometer used in the experiment, called a *Mach-Zehnder interferometer*, the light reaches the detectors via two different, spatially separated paths (analogous to the two slits in the double-slit experiment). Like in the first part of the experiment, the light is split at the first beam splitter. It is deflected by the mirrors and arrives at the second beam splitter, where the two partial beams are recombined. Only the combined beams are registered by the detectors.

There are two paths from the first beam splitter to the detector. Under these circumstances, interference occurs in classical optics, provided that the appropriate requirements for coherence, proper alignment, etc., are met. To demonstrate interference, the experimenters created a path difference between the two paths by slightly shifting one of the deflection mirrors, thereby changing the path length in the corresponding interferometer arm. The resulting count rates for both detectors depended on the path difference. Figure 3.9(b) shows the count rate of detector 2. The constructive and destructive interference interplay as a function of path difference is clearly visible. Detector 1 shows a similar pattern in which maxima and minima are reversed.

Explanation with the basic rules

The results of the second part of the experiment can likewise be explained by the basic rules. Since the second beam splitter mixes the two partial beams, there are two indistinguishable possibilities for the experimental result "detector 2 responds", namely via path 1 or via path 2. According to basic rule 2, interference occurs when the path length in the interferometer arms is varied.

At first glance, the results of the two parts of the experiment seem to contradict each other: The occurrence of interference is a clear indication that the photon "takes both paths at once", whereas in the anticorrelation experiment, the photon is only found on exactly one of the two paths during a measurement. We have already explained the latter with rule 3, which describes the special character of measurement in quantum physics. It is also possible to argue with rule 4 (complementarity): The "which-path" information obtained with the detectors in the first part and the interference observed in the second part are mutually exclusive. You can do an experiment in which you observe

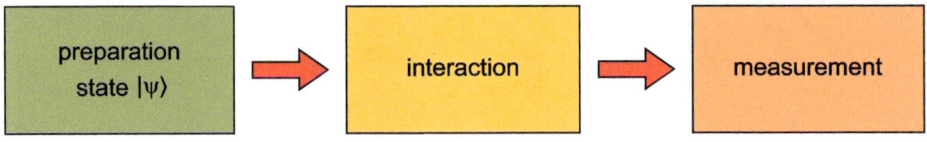

Figure 3.10: Scheme of an experiment in quantum physics.

interference or one in which you get "which-path" information, but it is not possible to get both in the same experiment.

3.4 States and measurements

With the mathematical formalism of quantum mechanics, two main types of predictions can be made:

1. Statements about the possible outcomes of a measurement. Quantization means that only a discrete set of possible values is found when measuring an observable. The formalism allows us to predict these possible measurement results (e. g., the possible energy values in an atom).
2. Statements about the probability distributions of measured values (such as the detector click rates in the experiment just discussed).

Figure 3.10 shows a general scheme that can be used to describe experiments in quantum physics. To perform experiments in a controlled way, quantum objects are first *prepared* in a certain way (e. g., photons with a certain direction and wavelength, atoms in a certain energy level, or qubits in a defined initial state). The quantum objects thus prepared are described by a state $|\psi\rangle$. Usually, an *interaction* follows (in which certain operations are performed on the quantum objects). In the end, a *measurement* of certain physical variables (often called *observables*) is performed. In the formalism of quantum mechanics, these three phases of an experiment are associated with different theoretical elements.

As already discussed in connection with rule 1, essential statements of quantum physics are of a statistical nature (cf. point 2 in the enumeration above). Thus, for a comparison between theoretical prediction and experiment, it is necessary to repeat an experiment very often. To describe this theoretically, *ensembles of identically prepared quantum objects* are considered. The quantum objects of such an ensemble are initially in the same state, and subsequently, the same manipulations and measurements are performed on them. State preparation is the systematic creation of such ensembles.

---

i *An example from quantum computing:* If a calculation on a quantum computer is carried out, it is not executed only once. A whole series of runs takes place, for example, $2^{10} = 1024$ (cf. Fig. 6.7 on p. 147). The reason for this is not only the notorious noise of currently available quantum computers but also the obtaining of

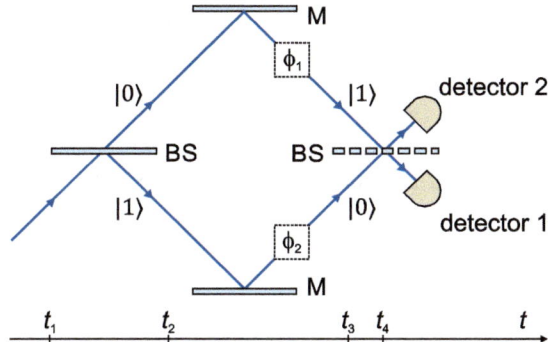

Figure 3.11: Modified representation of the experiment by Grangier, Roger, and Aspect.

information about the corresponding probability distributions in this way. In each run, the scheme of Fig. 3.10 is executed: The qubits are prepared in the initial state $|0\rangle$, bit operations are performed, and at the end, their value is measured. The result of many runs will be a histogram like shown in Fig. 6.7.

States and superpositions

Figure 3.11 shows the experiment of Grangier, Roger, and Aspect in a view rotated by 45°. The second beam splitter is shown with dashed lines. It is missing in the first part of the experiment and is only inserted in the second part. The two boxes marked ϕ are phase shifters. They are only used in the second part of the experiment and symbolically represent the path difference between the two partial beams created by shifting the mirrors. As in classical optics, phase jumps occur during reflection at the mirrors, but since they remain constant throughout the experiment, they are irrelevant here. Interference experiments only detect phase *differences*.

For a first introduction to the formalism, we consider an extremely reduced world [22, 23] in which, as in Fig. 3.11, there are only single photons running diagonally upwards or downwards (more precisely, we restrict ourselves to single-photon states in which there are only two values of the momentum vector $\vec{p} = \hbar\vec{k}$). These two states can be viewed as physical realizations of the qubit basis states $|0\rangle$ and $|1\rangle$. In addition, our world contains beam splitters, mirrors, phase shifters, and detectors. With these ingredients, we will be able to formally describe the part of the quantum mechanical formalism that is relevant to qubits.

Qubits, with only two basis states, are mathematically the simplest conceivable quantum systems. They are described by two-component vectors whose components can have complex values.

Qubit states: Qubits are described by two-component complex-valued vectors:

$$|\psi\rangle = \begin{pmatrix} \alpha \\ \beta \end{pmatrix}. \tag{3.3}$$

Two states $|0\rangle$ and $|1\rangle$ are singled out and referred to as the *computational basis*:

$$|0\rangle = \begin{pmatrix} 1 \\ 0 \end{pmatrix} \quad \text{and} \quad |1\rangle = \begin{pmatrix} 0 \\ 1 \end{pmatrix}. \tag{3.4}$$

These two states completely span the space of possible states. All other states can be composed from them by superposition:

$$\begin{pmatrix} \alpha \\ \beta \end{pmatrix} = \alpha |0\rangle + \beta |1\rangle . \tag{3.5}$$

The two states $|0\rangle$ and $|1\rangle$ are by no means the only possible basis. In the case of qubits, it is just particularly obvious to use these vectors. In general, any two other orthogonal vectors spanning the same state space form a completely equivalent basis. This fact is exploited in particular in quantum cryptography, which makes use of different bases for the polarization states of light.

The *superposition principle* describes the fact that the superposition of two physically possible states yields a third physically possible state. In quantum physics, the superposition principle has the same form as in all linear wave theories but with the more abstract quantum mechanical states instead of "real" waves.

> *Superposition principle*: If $|\psi_1\rangle$ and $|\psi_2\rangle$ are physically possible states of a quantum system, then all superposition states
>
> $$|\psi\rangle = \alpha |\psi_1\rangle + \beta |\psi_2\rangle \tag{3.6}$$
>
> are also physically possible states of the system.

⚡ The common parlance: "A qubit cannot only take the values 0 and 1, but also all values in between" can now be made more precise. Mathematically, we are talking about a superposition in a two-dimensional complex vector space. The superposition state (3.6) is described by the direction of a vector in this space (we will see in a moment that its length does not matter).

Scalar product and Dirac notation

Another mathematical structure is defined for quantum mechanical state vectors: the *scalar product*. Its definition is taken from linear algebra, where the scalar product $\langle x, y \rangle$ of two n-dimensional complex-valued vectors $x = (x_1, \ldots, x_n)$ and $y = (y_1, \ldots, y_n)$ is given by

$$\langle x, y \rangle = \sum_{i=0}^{n} x_i^* y_i \tag{3.7}$$

The asterisk means complex conjugation.

In quantum mechanics, a very powerful notation has been established, which goes back to Paul Dirac and greatly simplifies symbolic calculations with state vectors. It distinguishes between "ket vectors" $|\psi\rangle$, which are defined as column vectors as in Eq. (3.3), and "bra vectors" $\langle\psi|$. These are row vectors that result from the ket vectors by complex conjugation and swapping of rows and columns (transposition):

$$\text{ket vector:} \quad |\psi\rangle = \begin{pmatrix} \alpha \\ \beta \end{pmatrix}; \quad \text{bra vector:} \quad \langle\psi| = (\alpha^* \quad \beta^*). \tag{3.8}$$

With this *bra-ket notation* (also called *Dirac notation*), the scalar product can be written by "plugging" bras and kets together to form a "bracket". It can be calculated according to the rules of vector multiplication from linear algebra.

Scalar product: The scalar product of two state vectors $|\psi_1\rangle = \begin{pmatrix} a \\ \beta \end{pmatrix}$ and $|\psi_2\rangle = \begin{pmatrix} \gamma \\ \delta \end{pmatrix}$ is given by:

$$\langle\psi_1|\psi_2\rangle = (\alpha^* \quad \beta^*)\begin{pmatrix} \gamma \\ \delta \end{pmatrix} = \alpha^*\gamma + \beta^*\delta. \tag{3.9}$$

The scalar product defined in this way has the property $\langle\psi_1|\psi_2\rangle = \langle\psi_2|\psi_1\rangle^*$. Two state vectors $|\psi_1\rangle$ and $|\psi_2\rangle$ for which $\langle\psi_1|\psi_2\rangle = 0$ are *orthogonal* to each other. A state vector $|\psi\rangle$ whose scalar product with itself is 1, $\langle\psi|\psi\rangle = 1$, is called *normalized*.

Example: Show that the qubit states $|0\rangle$ and $|1\rangle$ are each normalized and orthogonal to each other.

Solution: To verify the normalization, we compute the scalar product of $|0\rangle$ with itself. The Dirac representation is:

$$|0\rangle = \begin{pmatrix} 1 \\ 0 \end{pmatrix} \quad \text{and} \quad \langle 0| = (1 \quad 0) \tag{3.10}$$

so that:

$$\langle 0|0\rangle = (1 \quad 0)\begin{pmatrix} 1 \\ 0 \end{pmatrix} = 1 \times 1 + 0 \times 0 = 1. \tag{3.11}$$

Thus, the vector $|0\rangle$ is normalized. Analogously, we find that $|1\rangle$ is also normalized, $\langle 1|1\rangle = 1$. The orthogonality of $|0\rangle$ and $|1\rangle$ follows from:

$$\langle 1|0\rangle = (0 \quad 1)\begin{pmatrix} 1 \\ 0 \end{pmatrix} = 0 \times 1 + 1 \times 0 = 0. \tag{3.12}$$

The two state vectors $|0\rangle$ and $|1\rangle$ thus form an *orthonormal system* for the two-dimensional state space of the qubit. An orthonormal system that completely spans the corresponding state space is called a *basis*.

Probabilities

The most important difference between quantum mechanics and classical physics is that the predictions of quantum mechanics are probability statements. Unlike in the statistical theories of classical physics, the occurrence of probabilities is not attributed to the

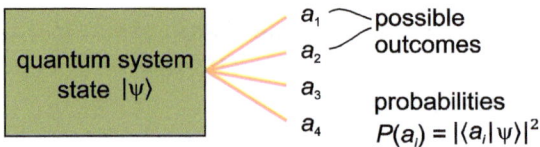

Figure 3.12: Illustration of Born's probability rule.

subjective ignorance of an observer (which could be avoided by better knowledge of the exact experimental conditions) but, as discussed in connection with basic rule 3, the probability is regarded as fundamental in quantum physics.

Probability is introduced into quantum mechanics by a postulate of its own: the probability formula, which goes back to Born (1926). It refers to a quantum system in the state $|\psi\rangle$, on which one makes measurements of the observable A. The possible measurement outcomes are $a_1, a_2, a_3, \ldots a_n$ (Fig. 3.12). To calculate the probability of finding the value a_j, the scalar product of $|\psi\rangle$ with the state vector $|a_j\rangle$ corresponding to the outcome a_j is taken. The probability $P(a_j)$ is then obtained by squaring.

> *Born's probability rule:* The probability of finding the outcome a_j when measuring the observable A on a system in state $|\psi\rangle$ is:
>
> $$P(a_j) = \left| \langle a_j | \psi \rangle \right|^2 \tag{3.13}$$

Example: Consider the qubit state $|\psi\rangle = \binom{\alpha}{\beta}$ and calculate the probabilities of finding the value 0 or 1 when a measurement in the computational basis ($|0\rangle$, $|1\rangle$) is performed.

Solution: To calculate the probabilities, we insert the states $|0\rangle$ and $|1\rangle$, which correspond to the measured values 0 and 1, into Born's probability formula:

$$P(0) = |\langle 0|\psi\rangle|^2 = \left| \begin{pmatrix} 1 & 0 \end{pmatrix} \begin{pmatrix} \alpha \\ \beta \end{pmatrix} \right|^2 = |\alpha|^2, \tag{3.14}$$

and likewise: $P(1) = |\langle 1|\psi\rangle|^2 = |\beta|^2$.

The two components of the state vector thus have a direct physical interpretation: the modulus squared $|\alpha|^2 = \alpha^* \alpha$ and $|\beta|^2 = \beta^* \beta$ is the probability of finding the value 0 or 1 upon a measurement in the computational basis. Therefore, the components of the state vector are called *probability amplitudes*.

Even though, we can only make statistical predictions about the outcome of measurements, it is clear that exactly one of the possible values will be found in each individual measurement. Thus, the sum of all probabilities must equal 1:

$$|\alpha|^2 + |\beta|^2 = 1. \tag{3.15}$$

This equation expresses the *normalization* condition for the components of the state vector. It is a general condition that also applies to higher-dimensional systems whose state vectors have more components. Geometrically, Eq. (3.15) can be interpreted as the square of the length of the state vector. The normalization condition states that in quantum mechanics, state vectors must always have length 1. Hence the statement already made in connection with Eq. (3.6) that only the direction of the state vector is relevant, not its length.

Eq. (3.13) also shows that the global phase factors $e^{i\phi}$ of a state vector are physically irrelevant because they drop out when the modulus is squared. The state vector $e^{i\phi}(\alpha |\psi_1\rangle + \beta |\psi_2\rangle)$ is equivalent in all respects to Eq. (3.6); all physical predictions are the same. Global phase factors are therefore generally ignored.

Application of the formalism

To illustrate the formalism with a concrete example, let us use it to describe the experiment of Grangier et al. We start with the first part, the anticoincidence experiment. According to the scheme in Fig. 3.10, we identify three phases of the experiment:

1. *Preparation:* As described on p. 48, the controlled preparation of single photons was the most difficult part of the experiment. We will ignore this experimental challenge and consider a source that emits single photons. In our reduced world, there are only two types of single photons: diagonally upwards (we denote this state by $|0\rangle$) and diagonally downwards (denoted by $|1\rangle$). We assume that the initial state is $|\psi(t_1)\rangle = |0\rangle$. Thus, the source prepares photons in the state $|0\rangle$, running diagonally upwards as shown in Fig. 3.11.

2. *Interaction:* The beam splitter puts the photons into a superposition of $|0\rangle$ and $|1\rangle$. After they have passed the beam splitter, at time t_2, their state is

$$|\psi(t_2)\rangle = t |0\rangle + r |1\rangle , \tag{3.16}$$

    where the coefficients t and r stand for "transmitted" and "reflected". For a 50/50 beam splitter, which transmits and reflects the beam in equal parts, t and r have the same magnitude.

    The two mirrors reflect the photons, swapping $|0\rangle$ and $|1\rangle$:

$$|\psi(t_3)\rangle = t |1\rangle + r |0\rangle . \tag{3.17}$$

    As in classical optics, the mirrors cause a phase jump of π on both partial beams, i. e., an additional factor $e^{i\pi} = -1$ in the state vector. As mentioned before, this global phase factor has no effect on the experimental results, so we ignore it.

3. *Measurement:* With the two detectors, we perform a measurement of the observable "diagonally upwards" or "diagonally downwards" (Fig. 3.11). According to basic rule 3, the measurement will give a definite result, i. e. exactly one of the two possible outcomes will be found. One of the detectors will be excited, but not both.

This is the anticoincidence already discussed above, which was confirmed in the experiment.

The measurement results are encoded with 1 and 0; their probabilities can be calculated according to Born's probability formula (3.13):

$$P(1) = |t|^2, \quad P(0) = |r|^2. \tag{3.18}$$

With probability $|t|^2$ a transmitted photon is measured, with probability $|r|^2$ a reflected one. Hence the terms transmission coefficient and reflection coefficient for t and r. For a 50/50 beam splitter, $|t|^2 = |r|^2 = \frac{1}{2}$.

Example: Check the expressions for the probabilities in Eq. (3.18).

Solution: We consider the state $\psi(t_3)$ and calculate the probabilities using Born's formula (3.13):

$$P(1) = \left| \langle 1|\psi(t_3) \rangle \right|^2 = \left| t \underbrace{\langle 1|1 \rangle}_{=1} + r \underbrace{\langle 1|0 \rangle}_{=0} \right|^2 = |t|^2, \tag{3.19}$$

$$P(0) = \left| \langle 0|\psi(t_3) \rangle \right|^2 = \left| t \underbrace{\langle 0|1 \rangle}_{=0} + r \underbrace{\langle 0|0 \rangle}_{=1} \right|^2 = |r|^2. \tag{3.20}$$

The outcome 1 belongs to the transmitted component $|1\rangle$ (in Fig. 3.11 running diagonally downwards at time t_3); the outcome 0 belongs to the reflected component $|0\rangle$ (running diagonally upwards at time t_3).

Description of the interference experiment

Let us consider the second part of the experiment. The second beam splitter is now inserted, creating a path difference between the two partial beams.

Before the photons reach the second beam splitter, their state is the same as in Eq. (3.17). We take into account the phases $e^{i\phi_1}$ and $e^{i\phi_2}$ on the two paths, although in the end, only the phase difference will matter. With an index 1, we indicate that the transmission and reflection coefficients refer to beam splitter 1:

$$|\psi(t_3)\rangle = t_1 e^{i\phi_1} |1\rangle + r_1 e^{i\phi_2} |0\rangle . \tag{3.21}$$

The two partial beams arrive at the second beam splitter, where both are partially reflected and partially transmitted. The effect of the second beam splitter on the states $|0\rangle$ and $|1\rangle$ is analogous to Eq. (3.16):

$$|0\rangle \rightarrow t_2 |0\rangle + r_2 |1\rangle ,$$
$$|1\rangle \rightarrow r_2 |0\rangle + t_2 |1\rangle . \tag{3.22}$$

The state at time t_4, after the photons have passed the second beam splitter, is obtained by substituting these expressions into Eq. (3.21):

$$|\psi(t_4)\rangle = (t_1 r_2 e^{i\phi_1} + r_1 t_2 e^{i\phi_2}) |0\rangle + (t_1 t_2 e^{i\phi_1} + r_1 r_2 e^{i\phi_2}) |1\rangle . \tag{3.23}$$

Alternatively, the state can also be constructed by tracing the possible paths in Fig. 3.11 and "collecting" the respective factors.

To show the occurrence of interference, we calculate the detector count rates as a function of the phase difference $\Delta\phi = \phi_1 - \phi_2$. We restrict ourselves to 50/50 beamsplitters for which $|t|^2 = |r|^2 = \frac{1}{2}$. Specifically, we choose $t_1 = t_2 = 1/\sqrt{2}$ and $r_1 = r_2 = i/\sqrt{2}$ (see the info box below for justification). Eq. (3.23) becomes:

$$|\psi(t_4)\rangle = \frac{i}{2}(e^{i\phi_1} + e^{i\phi_2})\,|0\rangle + \frac{1}{2}(e^{i\phi_1} - e^{i\phi_2})\,|1\rangle. \tag{3.24}$$

We take a common factor $\exp(i(\phi_1 + \phi_2)/2)$ out of the parentheses, use $e^{i\phi_1} = e^{i(\phi_1+\phi_2)/2}e^{i(\phi_1-\phi_2)/2}$, substitute $\phi_1 - \phi_2 = \Delta\phi$, use Euler's formula $e^{iz} = \cos(z) + i\sin(z)$, and omit the global phase factor $ie^{i(\phi_1+\phi_2)/2}$. The result is:

$$|\psi(t_4)\rangle = \cos\left(\frac{1}{2}\Delta\phi\right)|0\rangle + \sin\left(\frac{1}{2}\Delta\phi\right)|1\rangle. \tag{3.25}$$

As in Eq. (3.18), we can now calculate the probabilities of finding the photon at detector 1 (outcome 1) or detector 2 (outcome 0):

$$P(1) = \sin^2\left(\frac{1}{2}\Delta\phi\right), \quad P(0) = \cos^2\left(\frac{1}{2}\Delta\phi\right). \tag{3.26}$$

The detection probabilities change periodically with the path difference. The count rates show the typical interplay of constructive and destructive interference (cf. Fig. 3.9). Interference maxima and minima occur periodically when $\Delta\phi$ is varied. The excitation probabilities for the two detectors are reciprocal: the maximum probability of detector 1 means the minimum probability of detector 2 and vice versa. For $\Delta\phi = 0$ (equal path lengths in both arms), $P(1) = 0$ and $P(0) = 1$: only the upper detector in Fig. 3.11 responds in this case.

Phase shifts at a beam splitter

The simplest example of a beam splitter is familiar from everyday life: If you shine a flashlight diagonally on a glass plate, part of the light is transmitted, and part is reflected. This effect can often be observed on window panes.

Beam splitter plates work exactly in this way: they are thin glass plates coated on one side to achieve the desired reflectance (Fig. 3.13 left). Another type of beam splitter consists of two rectangular glass prisms glued together to form a cube (fig. 3.13 right); one side of the contact surface is coated.

From a theoretical point of view, beam splitters are not the simplest devices (cf., e. g., [24, 25]). The coefficients r and t in the formula (3.16) are generally complex, and only their magnitude is determined by the reflection and transmission properties of the material. The phases depend on the particular design, i. e., on the location of the reflective layers, so they differ for different types of beam splitters.

The following general statement can be proven with an energy conservation argument [23]: The phase difference between r and t must always be $\frac{\pi}{2}$. Because of $e^{i\pi/2} = i$, this requirement is fulfilled by the expressions $t = 1/\sqrt{2}$, and $r = i/\sqrt{2}$ we used above for the 50/50 beam splitter.

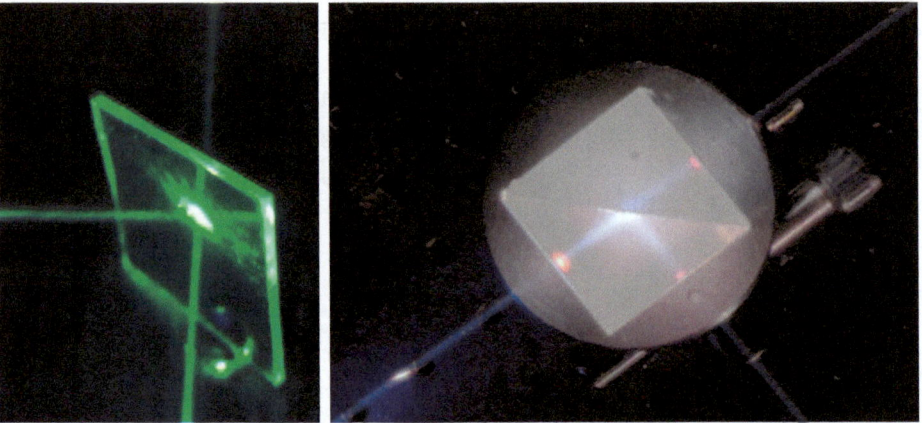

Figure 3.13: Plate and cube beam splitters.

3.5 Time evolution and qubit operations

Manipulations on qubits are described by *unitary matrices*. These are matrices for which

$$U^\dagger U = \mathbb{1}. \tag{3.27}$$

$U^\dagger$ (pronounced: „U dagger") denotes the *adjoint* of the matrix U. It is obtained from U by swapping rows and columns (transposition) and complex conjugation of all entries. $\mathbb{1}$ is the unit matrix.

Example: Show that the matrix

$$U_{BS} = \frac{1}{\sqrt{2}} \begin{pmatrix} 1 & i \\ i & 1 \end{pmatrix} \tag{3.28}$$

is unitary.

Solution: First, we construct the adjoint matrix. Since U is symmetric, it remains unchanged when swapping rows and columns. In the complex conjugate, i is replaced by $-i$ so that:

$$U_{BS}^\dagger = \frac{1}{\sqrt{2}} \begin{pmatrix} 1 & -i \\ -i & 1 \end{pmatrix}. \tag{3.29}$$

Now we perform the matrix multiplication:

$$U_{BS}^\dagger U_{BS} = \frac{1}{2} \begin{pmatrix} 1 & -i \\ -i & 1 \end{pmatrix} \begin{pmatrix} 1 & i \\ i & 1 \end{pmatrix} = \frac{1}{2} \begin{pmatrix} 2 & 0 \\ 0 & 2 \end{pmatrix} = \mathbb{1}. \tag{3.30}$$

This shows the unitarity of U_{BS}.

Example: Show that $U^\dagger U = \mathbb{1}$ also implies $UU^\dagger = \mathbb{1}$.

Solution: We multiply $U^\dagger$ from the left by $U^\dagger U = \mathbb{1}$:

$$U^\dagger = \mathbb{1} U^\dagger = \left(U^\dagger U\right)U^\dagger = U^\dagger\left(UU^\dagger\right). \tag{3.31}$$

For the left and right sides of the equation to be equal, $UU^\dagger = \mathbb{1}$ must hold. For completeness, it should be noted that in infinite-dimensional vector spaces, one can construct cases where the relation is no longer valid.

Time evolution

The *time evolution* of states is described in quantum mechanics by the *Schrödinger equation*. This is a differential equation whose formal solution can be written as

$$|\psi(t_1)\rangle = U\,|\psi(t_0)\rangle\,, \tag{3.32}$$

where U is a unitary matrix. Usually, one speaks of a unitary *operator* because U performs an "operation" on $|\psi(t_0)\rangle$. In quantum technologies, the operator U appears in two equivalent roles:

1. U as a *time evolution operator* $U = e^{-iH(t_1-t_0)/\hbar}$, which describes the continuous time evolution of the system between times t_0 and t_1. Here, H is an operator that specifies the total energy and thus characterizes the system. It is called the *Hamilton operator*. In this book, we will rarely use this role of U because in concrete applications the following form is more useful.

2. U is a well-defined operation on quantum systems that transforms a given initial state into a desired final state. The prototypical example of this description is the *quantum gates* used to perform controlled operations on qubits in quantum computers. The action of microwave or laser pulses interacting with the qubits is described by a unitary operator U characterizing the gate. In this approach, a more compact description in terms of "standard operations" is chosen instead of a continuous description of the time evolution of the system.

Unitarity of the time evolution, and, in particular, of the operators for gate operations, is necessary to preserve the normalization of $|\psi\rangle$ and thus to guarantee that the sum over all probabilities equals 1 at each time (cf. Eq. (3.15)). This can be shown as follows: We assume that the state at time t_0 is normalized, $\langle\psi(t_0)|\psi(t_0)\rangle = 1$. At a later time t_1, we have:

$$\langle\psi(t_1)|\psi(t_1)\rangle = \langle\psi(t_0)|U^\dagger U|\psi(t_0)\rangle\,. \tag{3.33}$$

This is equal to 1 exactly if $U^\dagger U = \mathbb{1}$, i. e. if U is unitary.

Example: Show that the operator U_{BS} from Eq. (3.28) describes the action of the symmetric beam splitter considered in the previous section:

$$U_{BS} = \begin{pmatrix} t & r \\ r & t \end{pmatrix} = \frac{1}{\sqrt{2}} \begin{pmatrix} 1 & i \\ i & 1 \end{pmatrix}. \tag{3.34}$$

Solution: We apply U_{BS} to the states $|0\rangle$ (photon traveling diagonally upwards) and $|1\rangle$ (photon traveling diagonally downwards) and compare with Eq. (3.22), which we used to describe the action of the beam splitter. We find:

$$U_{BS} |0\rangle = \frac{1}{\sqrt{2}} \begin{pmatrix} 1 & i \\ i & 1 \end{pmatrix} \begin{pmatrix} 1 \\ 0 \end{pmatrix} = \begin{pmatrix} 1 \\ i \end{pmatrix}, \tag{3.35}$$

$$U_{BS} |1\rangle = \frac{1}{\sqrt{2}} \begin{pmatrix} 1 & i \\ i & 1 \end{pmatrix} \begin{pmatrix} 0 \\ 1 \end{pmatrix} = \begin{pmatrix} i \\ 1 \end{pmatrix}. \tag{3.36}$$

Overall, therefore:

$$U_{BS} |0\rangle = \frac{1}{\sqrt{2}} (1 |0\rangle + i |1\rangle),$$

$$U_{BS} |1\rangle = \frac{1}{\sqrt{2}} (i |0\rangle + 1 |1\rangle). \tag{3.37}$$

This corresponds to Eq. (3.22) with $t = 1/\sqrt{2}$ and $r = i/\sqrt{2}$. The unitary matrix U_{BS} thus describes the action of a symmetric beam splitter.

Example: Construct the unitary matrix describing the action of the mirrors in Fig. 3.11.

Solution: A mirror in Fig. 3.11 turns photons running diagonally upwards into photons running diagonally downwards and vice versa. It exchanges the states $|0\rangle$ and $|1\rangle$. Additionally, it generates a phase jump of $\frac{\pi}{2}$, which we will ignore as before. The exchange of $|0\rangle$ and $|1\rangle$ is achieved by means of the matrix

$$U_M = \begin{pmatrix} 0 & 1 \\ 1 & 0 \end{pmatrix}. \tag{3.38}$$

The matrix is its own adjoint. Its unitarity can be checked by matrix multiplication. On p. 144, we will encounter it again as a quantum gate. It describes the *Pauli-X-gate*, the quantum mechanical generalization of the classical NOT gate.

Using the bra-ket formalism: outer product

The bra-ket formalism is a very powerful tool for symbolic calculations. In particular, the *outer product* is an elegant way to represent operators. It represents a matrix A by the matrix elements A_{ij} and the unit vectors $|u_i\rangle$ of the underlying basis:

$$A = \sum_{ij} A_{ij} |u_i\rangle \langle u_j|. \tag{3.39}$$

When an operator written in this form is applied to states, scalar products result, which can be evaluated as usual. The same applies when several operators are used in succession. Without having to go into the mathematical details, you can rely on the Dirac notation to guide you in the correct application of the formalism.

The outer product can be used to obtain a useful representation of the unit matrix $\mathbb{1}$ using the basis vectors $|u_i\rangle$ of a complete orthonormal system. We write

$$\mathbb{1} = \sum_{ij} \mathbb{1}_{ij} |u_i\rangle \langle u_j| \tag{3.40}$$

and immediately obtain the following equation, called the *completeness relation*:

$$\mathbb{1} = \sum_{i} |u_i\rangle \langle u_i| . \tag{3.41}$$

Example: Show that the matrix (3.38) describing the action of a mirror (as well as the Pauli-*X* gate) can be expressed in the form

$$U_M = |0\rangle \langle 1| + |1\rangle \langle 0| . \tag{3.42}$$

Solution: To prove Eq. (3.42), we apply the operator U_M to the state vector $|0\rangle$. This results in:

$$U_M |0\rangle = \big[|0\rangle \langle 1| + |1\rangle \langle 0|\big] |0\rangle = |0\rangle \langle 1|0\rangle + |1\rangle \langle 0|0\rangle . \tag{3.43}$$

Formal expansion of the square brackets has led to the scalar products $\langle 1|0\rangle = 0$ and $\langle 0|0\rangle = 1$. Substituting back, we obtain as expected:

$$U_M |0\rangle = |1\rangle . \tag{3.44}$$

Analogously, applying U_M to the state vector $|1\rangle$ yields:

$$U_M |1\rangle = \big[|0\rangle \langle 1| + |1\rangle \langle 0|\big] |1\rangle = |0\rangle \underbrace{\langle 1|1\rangle}_{=1} + |1\rangle \underbrace{\langle 0|1\rangle}_{=0} = |0\rangle . \tag{3.45}$$

Thus, the operator U_M represented by Eq. (3.42) acts on the state vectors in the same way as the matrix (3.38): it exchanges $|0\rangle$ and $|1\rangle$.

3.6 Observables and measurements

In the scheme of a quantum physics experiment shown in Fig. 3.10, the *measurement* is the third step. We have already discussed the inherently statistical character of a measurement in quantum physics in connection with basic rule 3. When measurement of an observable A on a quantum system in the state $|\psi\rangle$ is performed, in general, only probability statements for the distribution of the measurement results can be made. The probabilities for the possible measured values of A can be calculated with Born's probability formula (3.13).

The quantum mechanical formalism can also be used to calculate the possible outcomes of a measurement of an observable A. Quantum physics owes its name to the fact that for some observables, not all, but only certain values are possible. The prediction

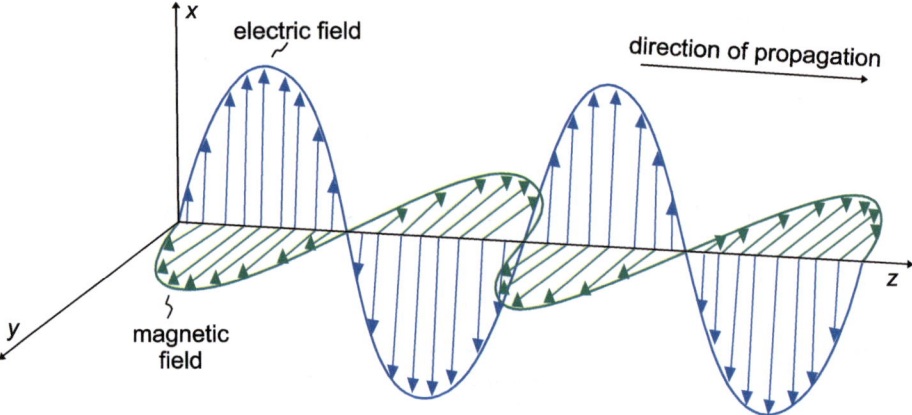

Figure 3.14: Linear polarization of light in the classical wave model.

of possible measured values with the formalism of quantum mechanics was historically of great importance. In the 1920s, it was a great triumph for quantum mechanics that it became possible to predict the quantized values of the energy of atoms, thus explaining the discrete nature of spectral lines.

Polarization of light

In the wave model of classical physics, the *polarization* of light can be regarded as the oscillation direction of the electric field. The electric field vector always oscillates in a plane perpendicular to the direction of propagation of the light. In the simplest case, the field vector oscillates linearly in a certain direction. This is called *linear polarization* (Fig. 3.14). However, more complicated patterns like circular or elliptical polarization are also possible (and are used not only in scientific applications but also for 3D cinema). In everyday life, we usually deal with unpolarized light, which we can imagine as a mixture of photons of all polarization directions without fixed phase relations.

We cannot perceive the polarization of light. With the naked eye, we cannot tell whether light is polarized or unpolarized, nor can we tell the direction of polarization. We have to rely on instruments to measure polarization. The easiest way to do this is to use inexpensive sheet polarizers. They allow only a certain linear polarization component of the light to pass through and absorb the component perpendicular to it. More sophisticated devices are *polarizing beam splitters* (PBS).

ℹ **Polarizing beam splitters**

Polarizing beam splitters typically consist of two prisms glued together to form a cube (Fig. 3.15). The contact surface is provided with a coating that determines the transmission and reflection properties of the beam splitter.

As with a normal beam splitter, incident light is split into a transmitted and a reflected part. Unlike a normal beam splitter, the light coming from the two outputs is linearly polarized (arrows in Fig. 3.15). The

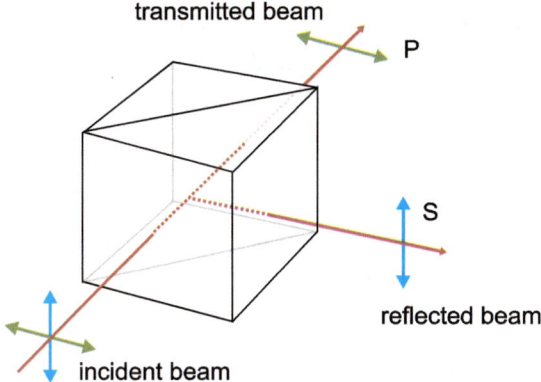

transmitted beam

P

S

reflected beam

incident beam

Figure 3.15: Polarizing beam splitter (cf. also Fig. 3.13).

behavior of a polarizing beam splitter can be demonstrated with a laser pointer, whose light is linearly polarized in many models. If the laser pointer is rotated around its axis while illuminating the beam splitter, the polarization direction of the incident light changes and, as a consequence, the intensity of the light emitted from the two outputs of the beam splitter also changes: for a certain polarization direction, light comes only from one output; with further rotation around 90°, light comes only from the other output.

The following properties of a polarizing beam splitter are relevant in the context of quantum mechanical measurement:
1. If horizontally polarized light hits the polarizing beams splitter, it is completely transmitted (green arrows in Fig. 3.15). No light is reflected.
2. If vertically polarized light hits the polarizing beam splitter, it is completely reflected (blue arrows in Fig. 3.15). Nothing is transmitted.
3. If the light hitting the polarizing beam splitter is obliquely polarized (at an angle a to the horizontal), it is split into linearly polarized partial beams. The intensity of reflected and transmitted partial beam depends on the angle a (it is proportional to $\cos^2 a$ and $\sin^2 a$, respectively). The transmitted part of the beam is fully horizontally polarized, and the reflected part is fully vertically polarized. Thus, a polarizing beam splitter can be used not only for detection but also for the systematic preparation of polarized light.

We are considering ideal beam splitters and sources that emit fully polarized light. For real devices, this is often only approximately correct. Then the intensities do not vary between 0 % and 100 %, but between "brighter" and "darker". The labels S and P shown in Fig. 3.15 are customary in applied optics. They denote the polarization of the outgoing light, which is either perpendicular (S) to the plane formed by the incident ray and the normal to the reflecting surface or parallel (P) to this plane. In quantum technologies, the terms H and V are more common, referring not to the beam splitter but to the orientation with respect to the laboratory bench.

We now move from classical optics to quantum physics and repeat the laser pointer experiment with linearly polarized single photons. The polarization is prepared, for example, by using another polarizing beam splitter. With this setup, we measure the observable "polarization in the H/V direction". As expected, horizontally polarized photons are completely transmitted, and vertically polarized photons are completely reflected.

In the case of obliquely polarized photons, the peculiarities of the quantum-mechanical measurement become apparent once again. According to basic rule 3, each measurement yields a definite value of the measured quantity. So also here: An incident photon is either transmitted as a whole (and detected at the P output) or reflected as a whole (and detected at the S output). The detection frequencies vary with the angle of incidence. Furthermore, all photons found at the P output are horizontally polarized; all photons found at the S output are vertically polarized. In the following, we will describe this experimental result using the quantum mechanical formalism.

Polarization states

As the basis states for the quantum mechanical description of polarization, we choose horizontal and vertical polarization. They are described by the following state vectors:

$$|H\rangle = \begin{pmatrix} 1 \\ 0 \end{pmatrix} \quad \text{and} \quad |V\rangle = \begin{pmatrix} 0 \\ 1 \end{pmatrix}. \tag{3.46}$$

Using these basis states, we can form arbitrary superpositions of polarization states. In particular, photons polarized at the angle α with respect to the horizontal are represented by the state

$$|Z\rangle = \cos\alpha\, |H\rangle + \sin\alpha\, |V\rangle. \tag{3.47}$$

If the previously described measurement of the observable "polarization in H/V direction" is performed on photons polarized with angle α, then the corresponding probabilities are given by Born's formula (3.13):

$$P(H) = \left| \langle H|Z \rangle \right|^2 = \cos^2\alpha,$$
$$P(V) = \left| \langle V|Z \rangle \right|^2 = \sin^2\alpha. \tag{3.48}$$

These probabilities indicate the relative frequencies with which H- and V-polarized photons will be detected at the corresponding outputs if the experiment is repeated many times.

In classical optics, the corresponding relation has long been known as *Malus' law*. Figure 3.16 shows the probability $P(H)$ as a function of α. Even more informative is the plot in the figure on the right, where $P(H)$ is plotted in a polar diagram as a function of α. It shows a "figure 8" distribution with the long axis aligned along the direction of the polarization measurement (i. e., the 0° direction for $P(H)$ and the 90° direction for $P(V)$).

We can generalize our polarization measuring device (the beam splitter cube together with single photon detectors) to the case of polarization directions other than H/V. To do so, we rotate the array in Fig. 3.15 by an angle β around the axis of the incident beam and thus measure the polarization in the β- or ($\beta - 90°$) direction. In quantum communication, a basis rotated 45° with respect to the H/V basis is often used. This

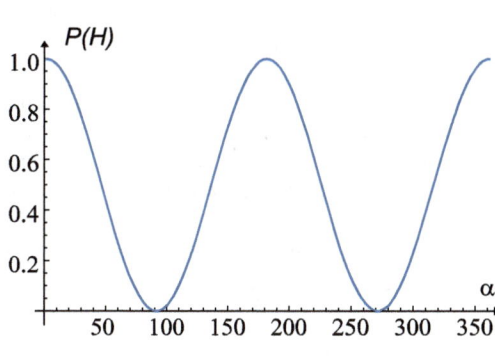
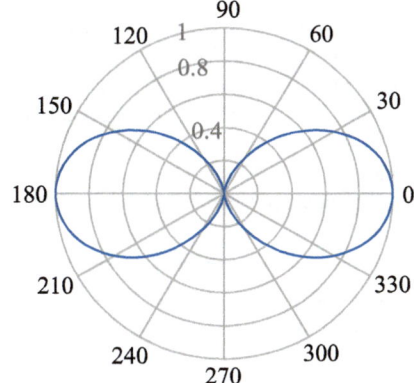

Figure 3.16: Probability for the result H as a function of the polarization angle a (drawn as a polar plot on the right).

basis is called the ±45° basis (or ± basis for short). The corresponding basis vectors are:

$$|+\rangle = \cos(45°)\,|H\rangle + \sin(45°)\,|V\rangle = \frac{1}{\sqrt{2}}(|H\rangle + |V\rangle) \tag{3.49}$$

and

$$|-\rangle = \frac{1}{\sqrt{2}}(|H\rangle - |V\rangle). \tag{3.50}$$

Observables and eigenvalues

In quantum mechanics, each observable is associated with an operator described by a *hermitian matrix*. Matrices are called hermitian if

$$A^\dagger = A \tag{3.51}$$

holds, i. e. if they are equal to their adjoints. Hermitian matrices and operators are therefore also called self-adjoint. They have the property that their *eigenvalues* are always real.

Eigenvalues and eigenvectors

The mathematical concept of *eigenvalues* and *eigenvectors* of a matrix is a prerequisite for the description of quantum mechanical observables and measurements. If for a given matrix A, there is a vector $|u_j\rangle$ and a number λ_j, satisfying the equation

$$A\,|u_j\rangle = \lambda_j\,|u_j\rangle\,, \tag{3.52}$$

then λ_j is an eigenvalue of A with the corresponding eigenvector $|u_j\rangle$. For a given matrix A, the *eigenvalue equation* (3.52) has solutions only for certain sets of numbers λ_j and their corresponding eigenvectors. Computing the eigenvalues of a matrix is a standard problem in linear algebra, and there is an extensive literature on the subject. In practice, the eigenvalues of an $n \times n$ matrix can be determined by solving a polynomial of degree n (characteristic polynomial).

An $n \times n$ matrix has n (not necessarily different) eigenvalues with corresponding eigenvectors. If the matrix is hermitian, the eigenvalues are real. The set of eigenvalues of A is called the *spectrum* of A. For infinite-dimensional operators (such as position or momentum), which are not described by matrices but by differential operators, the spectrum is often no longer discrete, and the eigenvalues take on continuous values.

Example: Show that the three so-called *Pauli matrices* are hermitian:

$$\sigma_x = \begin{pmatrix} 0 & 1 \\ 1 & 0 \end{pmatrix}, \quad \sigma_y = \begin{pmatrix} 0 & -i \\ i & 0 \end{pmatrix}, \quad \sigma_z = \begin{pmatrix} 1 & 0 \\ 0 & -1 \end{pmatrix}. \tag{3.53}$$

Solution: We need to form the adjoint of all three matrices, i. e. swap rows and columns and then take the complex conjugate. For σ_x and σ_z, we see immediately that the result is equal to the original matrix. For σ_y, we obtain similarly:

$$\sigma_y^\dagger = \left(\sigma_y^T\right)^* = \begin{pmatrix} 0 & i \\ -i & 0 \end{pmatrix}^* = \begin{pmatrix} 0 & -i \\ i & 0 \end{pmatrix} = \sigma_y. \tag{3.54}$$

Pauli matrices often appear in connection with observables that can take on exactly two values, such as the spin of electrons. Since they are not only hermitian but also unitary, they can also describe operations that are performed on quantum systems. We have already encountered the matrix σ_x in this role in the description of the mirror in Eq. (3.38). We will meet the Pauli matrices again as quantum gates in quantum computing.

In quantum mechanics, the eigenvalue equation (3.52) for an observable is of central importance because it can be used to predict the possible outcomes of a measurement of that observable. This is described by the following rule.

> *Measurement results in quantum mechanics:* The possible measured values of an observable are the eigenvalues of the associated operator.

We already know from basic rule 3 that in each measurement, a definite value of the measured variable is found. The question of what these possible outcomes are can be answered with the help of the eigenvalue equation: The possible results of a measurement are the eigenvalues λ_j of the respective operator. Values other than the eigenvalues are not found. This is the origin of quantization.

The probabilities for the possible outcomes are calculated with Born's probability formula (3.13). The states used in this formula to calculate the probability of finding the value λ_j are the corresponding eigenstates $|u_j\rangle$. Therefore, according to Eq. (3.13), the probability of finding the value λ_j when measuring a system in the state $|\psi\rangle$ is

$$P(\lambda_j) = |\langle u_j|\psi\rangle|^2. \tag{3.55}$$

Thus, via Born's probability formula, the elements of the formalism (operators, eigen-values, and eigenstates) are directly linked to experimental experience.

It follows from Eq. (3.55) that the eigenstates of an observable are those states in which the corresponding measurement result is found with probability 1 because if $|\psi\rangle = |u_j\rangle$, then according to Eq. (3.55) the probability for the associated eigenvalue is $P(\lambda_j) = 1$; it is zero for all other values. Thus, the eigenstates of an observable are the states in which one specific value is always found when that observable is measured. The measurement results do not scatter in this case.

Example: Show that the operator

$$\sigma_z = \begin{pmatrix} 1 & 0 \\ 0 & -1 \end{pmatrix} \tag{3.56}$$

can be used to describe the measurement with single photons at the polarizing beam splitter discussed on p. 65.

Solution: To mathematically describe the experiment, we encode the measurement result "photon found at output P" (horizontal polarization) with +1 and the measurement result "photon found at output S" (vertical polarization) with −1. These are the possible measurement values that the desired operator must provide as eigenvalues.

For classical light, the behavior at the beam splitter has been described on p. 65 for different directions of polarization. Analogously, for single photons, we obtain the following experimental results:
1.   For photons in the state $|H\rangle$, the measured value is always 1 (i. e., with probability 1).
2.   For photons in the state $|V\rangle$, the measured value is always −1.
3.   If the photons are in a superposition state, the corresponding probabilities are $\cos^2 \alpha$ and $\sin^2 \alpha$.

We conclude that the states $|H\rangle$ and $|V\rangle$ must be eigenstates of the operator we are looking for since they give the corresponding outcomes +1 and −1 with probability 1. We check if this is true for the matrix σ_z. Applying it to $|H\rangle$, we get:

$$\sigma_z |H\rangle = \begin{pmatrix} 1 & 0 \\ 0 & -1 \end{pmatrix} \begin{pmatrix} 1 \\ 0 \end{pmatrix} = \begin{pmatrix} 1 \\ 0 \end{pmatrix} = +1 |H\rangle . \tag{3.57}$$

The eigenvalue equation (3.52) is thus satisfied, and as desired, $|H\rangle$ is indeed an eigenvector of σ_z with eigenvalue +1. Similarly, it follows that

$$\sigma_z |V\rangle = -1 |V\rangle . \tag{3.58}$$

The state $|V\rangle$ is also an eigenstate of σ_z with eigenvalue −1. Thus, we have identified the two eigenstates and eigenvalues of σ_z. Using Born's rule, we obtain the probabilities for superposition states from $|H\rangle$ and $|V\rangle$, as already calculated in Eq. (3.48).

Expectation values

The *expectation value* $\langle A \rangle$ of an observable A describes the average of the measured values over many measurements. It is obtained, as in classical statistics, by summing the measured values weighted with the probabilities:

$$\langle A \rangle = \sum_j P(\lambda_j)\lambda_j = \sum_j |\langle u_j|\psi\rangle|^2 \lambda_j. \tag{3.59}$$

Using Dirac's notation, this can be written in a very compact way as

$$\langle A \rangle = \langle \psi|A|\psi\rangle. \tag{3.60}$$

Since the eigenvalues of A form a complete orthonormal system, we can insert the unity operator using Eq. (3.41) and write:

$$\langle \psi|A \mathbb{1}|\psi\rangle = \sum_j \langle \psi|A|u_j\rangle \langle u_j|\psi\rangle = \sum_j \langle \psi|u_j\rangle \langle u_j|\psi\rangle \lambda_j = \sum_j |\langle u_j|\psi\rangle|^2 \lambda_j.$$

In quantum technologies, one is less interested in averages than in traditional applications of quantum mechanics because single events play a much larger role here.

Using Eq. (3.41), we can derive another helpful relation. A hermitian matrix A has a complete system of eigenvectors $|u_j\rangle$ with eigenvalues λ_j. The eigenvalue equation is $A|u_j\rangle = \lambda_j|u_j\rangle$. With Eq. (3.41), the following representation of A in terms of the outer products of the eigenvectors can be shown:

$$A = \sum_{j=1}^{N} \lambda_j |u_j\rangle \langle u_j|. \tag{3.61}$$

Example: Show that Eq. (3.61) is indeed valid.

Solution: For the hermitian matrix A, the eigenvectors $|u_j\rangle$ form a complete system of basis vectors. Eq. (3.41), written in this basis, reads:

$$\mathbb{1} = \sum_j |u_j\rangle \langle u_j|. \tag{3.62}$$

We apply the operator A to both sides of the equation to show the validity of Eq. (3.61):

$$A \times \mathbb{1} = \sum_j A|u_j\rangle \langle u_j| = \sum_j \lambda_j |u_j\rangle \langle u_j|. \tag{3.63}$$

In the second step, the eigenvalue equation was used. Thus, the operator A is diagonal in its eigenvector basis and has its eigenvalues as diagonal elements.

3.7 Indeterminacy

From our classical experience, we are not used to encountering objects that just do not possess certain properties – for example, electrons that cannot be assigned a certain position within the atom or helium atoms that cannot be assigned a definite path in the double-slit experiment (cf. p. 43). The superposition of $|0\rangle$ and $|1\rangle$ in qubits is another example. Before the discovery of quantum mechanics, there was no reason to communicate about such states, so our language is not able to adequately describe them. When formulating statements about such states, special care is needed. The *Heisenberg indeterminacy relation* is a case where this becomes particularly clear.

The indeterminacy relation refers to *pairs of properties*. Most often, position and momentum are discussed. In quantum technologies, the polarization of photons is more important because it is used to encode information. Therefore, we will discuss this example.

We have seen that there are states for which the same value is always found when the observable A is measured. These are the eigenstates of A. For the polarizing beamsplitter, it is the states $|H\rangle$ and $|V\rangle$, where (assuming ideal conditions) the corresponding eigenvalue (+1 or −1) is always found. If one of the eigenstates is incident on the beam splitter, there is no scattering in the measured data: all photons exit at the same output. We can thus assign a definite value of H/V polarization to photons in one of these two eigenstates. On the other hand, photons in a superposition state of $|H\rangle$ and $|V\rangle$ have no defined H/V property; the H/V outcomes scatter when the measurement is repeated many times.

Heisenberg's indeterminacy relation answers the question under which circumstances it is possible to prepare a state for which the measured values of *two* observables do not scatter. In other words: Is it possible to prepare states so that both the measurement of observable A and the measurement of observable B always yield the same values? Quantum mechanics allows formulating a simple criterion for this: The corresponding operators have to commute. For the matrices A and B, the relation $AB = BA$ must hold. In this case, states can be prepared that are eigenstates of both A and B. Using the notation $[A, B] = AB - BA$ (commutator) we can write:

> It is possible to prepare states for which the measured values of two observables A and B both do not scatter if the associated operators commute: $[A, B] = 0$.

Different polarization bases

We have seen that there are exactly two states $|H\rangle$ and $|V\rangle$ that, when measured with the polarizing beam splitter in Fig. 3.15, lead to definite measurement outcomes without scattering. We call this measurement the *H/V measurement*; the two states $|H\rangle$ and $|V\rangle$ form the H/V basis.

We can define a new observable by rotating the beam splitter cube by 45° (Fig. 3.17). Like the original beam splitter, the rotated arrangement splits incident light into two partial beams. Due to the rotation, the transmitted photons now have a polarization of −45°, the reflected ones +45° with respect to the original H/V plane. The exit direction of the reflected photons is also rotated accordingly. We call this measurement the ± *measurement*.

In the context of the Heisenberg indeterminacy relation, we ask: Is it possible to prepare the two observables simultaneously? Are there states for which the outcomes of both the H/V measurement and the ± measurement do not scatter?

Since there are only two states with definite outcomes for the H/V measurement (the eigenstates discussed above), it is sufficient to consider these two states and check whether a ± measurement will also have

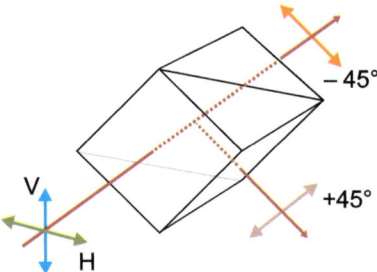

Figure 3.17: Polarization measurement with a polarizing beam splitter rotated by 45°.

a definite outcome. It turns out that this is not the case. On the contrary: The measured values are completely random; they show maximum scattering. For photons in the $|H\rangle$ state, the ± measurement gives a 50/50 probability of detection at the + output or at the − output. The same is true for photons in the $|V\rangle$ state.

How can this experimental result be represented in the formalism? We have already found in Eq. (3.49) the two states $|+\rangle$ and $|-\rangle$ that describe photons polarized by ±45°. We call these two states the ± basis. By substituting them into the eigenvalue equation, we can convince ourselves that they are eigenstates of the following operator, which describes the ± measurement:

$$\sigma_{45°} = \begin{pmatrix} \cos(2 \times 45°) & \sin(2 \times 45°) \\ \sin(2 \times 45°) & -\cos(2 \times 45°) \end{pmatrix} = \begin{pmatrix} 0 & 1 \\ 1 & 0 \end{pmatrix}. \tag{3.64}$$

It is the matrix σ_x of Eq. (3.53); the states $|+\rangle$ and $|-\rangle$ are its eigenstates with eigenvalues +1 and −1.

To check whether it is possible to prepare the H/V observable and the ± observable simultaneously, we compute the commutator of the two corresponding operators σ_z and σ_x:

$$[\sigma_z, \sigma_x] = \begin{pmatrix} 1 & 0 \\ 0 & -1 \end{pmatrix}\begin{pmatrix} 0 & 1 \\ 1 & 0 \end{pmatrix} - \begin{pmatrix} 0 & 1 \\ 1 & 0 \end{pmatrix}\begin{pmatrix} 1 & 0 \\ 0 & -1 \end{pmatrix} = \begin{pmatrix} 0 & 2 \\ -2 & 0 \end{pmatrix}. \tag{3.65}$$

The commutator is different from zero; the two operators do not commute. Thus, according to the indeterminacy relation, it is not possible to prepare both observables simultaneously. There are no states in which both the H/V observable and the ± observable have fixed values that appear with certainty in every measurement. Put another way, there are no quantum systems to which we can ascribe both the H/V property and the ± property at the same time. This fact is crucial for quantum communication. Here, the incompatibility of the two bases is exploited to prevent eavesdropping when exchanging cryptographic keys.

Example: Consider an ensemble of photons prepared in the state $|V\rangle$. Determine the probabilities for the two possible outcomes of a measurement in the ± basis.

Solution: We calculate the probabilities with Born's formula (3.13). With respect to the H/V basis, the states of the ± basis are as given in Eq. (3.49):

$$|+\rangle = \frac{1}{\sqrt{2}}\left(|H\rangle + |V\rangle\right), \tag{3.66}$$

$$|-\rangle = \frac{1}{\sqrt{2}}\left(|H\rangle - |V\rangle\right). \tag{3.67}$$

We assume that the photons incident on the beam splitter are prepared in the state $|V\rangle$. The probability for the outcome + according to Born's probability formula is:

$$P(+) = \left|\langle+|V\rangle\right|^2 = \frac{1}{2}\left|\underbrace{\langle H|V\rangle}_{=0} + \underbrace{\langle V|V\rangle}_{=1}\right|^2 = \frac{1}{2}. \tag{3.68}$$

The same result is found for the outcome −. As already mentioned above, half of the photons are found at the output +, and the other half at the output − in the measurement. It is not predictable (and completely random) at which of the two outputs an individual photon is found.

Heisenberg's indeterminacy relation makes a statement about the possibility of simultaneously preparing two observables with the smallest possible data scatter. Mathematically, the scatter of the measured values for the observable A can be characterized by the *standard deviation* ΔA. Each measurement of A yields a frequency distribution of the measured values. As in classical statistics, the width of the distribution can be described by the standard deviation. In quantum mechanics, ΔA can be determined by calculating the following expectation values:

$$(\Delta A)^2 = \langle A^2 \rangle - \langle A \rangle^2. \tag{3.69}$$

The definition of ΔA is analogous to classical statistics, except that instead of classical mean values, quantum mechanical expectation values are used (for notation, see Eq. (3.60)). In the same way, ΔB can be calculated for another observable B. ΔB describes the standard deviation of the frequency distribution of B measurements. The A measurements and the B measurements are performed independently on identically prepared ensembles of quantum objects. The Heisenberg indeterminacy relation gives a statement about the product of the two standard deviations.

> *Heisenberg's indeterminacy relation:* If we prepare a quantum system in the state $|\psi\rangle$ and perform measurements on two observables A and B, then the following inequality holds for the product of their standard deviations:
>
> $$\Delta A\,\Delta B \geq \langle\psi|[A, B]|\psi\rangle. \tag{3.70}$$

This statement generalizes the above rule about the simultaneous preparability of two observables. Since the formula is inequality, the product $\Delta A \times \Delta B$ can always be greater than the value on the right, depending on the concrete system and the type of measurement (it is always possible to make imprecise measurements that have large data scatter). On the other hand, the product can never be less than specified by Eq. (3.70).

Related indeterminacy relations

In the literature, one finds various, more or less closely related uses of the term "Heisenberg indeterminacy relation" (or "uncertainty relation"):

1. In the version formulated above, the measurements of A and B are performed independently of each other on ensembles of identically prepared quantum objects in the state $|\psi\rangle$. This interpretation is also called *preparation indeterminacy* (because it deals with the simultaneous preparability of two observables). Mathematically, this is reflected in Eq. (3.69), which corresponds to the definition of the standard deviation in classical statistics.

2. In the statements above, there is no mention of *simultaneous* measurements and their possible interference. Of course, it is also conceivable to build measuring devices that simultaneously measure two observables A and B. In 2004, Werner was able to derive a similar relationship that applies to the case of simultaneous measurements [26].

3. It is often said that the measurement of the observable A affects or perturbs the measurement of B. This statement refers to successive measurements of the same quantum objects ("perturbation by a measurement"). Historically, it is the situation analyzed by Heisenberg using thought experiments. Again, this is a legitimate issue that has recently been analyzed more carefully [27, 28]. The slogan "no measurement without perturbation" summarizes that in quantum mechanics, measuring a system without changing its state is generally not possible.

4. There are also more heuristic "Heisenberg relations". In many cases, these are examples of the relation $\Delta k \times \Delta x \geq 1$ between the spectral width and the spatial (or temporal) extension of wave trains, which is well known from the mathematical theory of Fourier transforms. It is valid in this form also in classical optics or acoustics and therefore cannot contain any specific statement about quantum phenomena.

3.8 Visualization with the Bloch sphere

The general state of a quantum system with two states $|0\rangle$ and $|1\rangle$ is the superposition state

$$|\psi\rangle = \alpha\,|0\rangle + \beta\,|1\rangle , \tag{3.71}$$

where α and β are complex numbers. Thus, the state is described by four real numbers (real part and imaginary part of α and β). The normalization condition $|\alpha|^2 + |\beta|^2 = 1$ reduces this number to three. The number of free parameters is reduced even further by the fact that the global phase of the state is physically irrelevant. Only the relative phase between $|0\rangle$ and $|1\rangle$ is relevant. We can therefore specify the state completely by specifying two real numbers. In a form that automatically satisfies the normalization condition, we write:

$$|\psi\rangle = \cos\frac{\theta}{2}\,|0\rangle + e^{i\phi}\sin\frac{\theta}{2}\,|1\rangle . \tag{3.72}$$

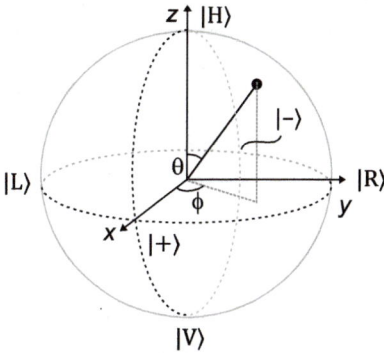

Figure 3.18: Bloch sphere for the polarization states of light.

We can identify the two parameters θ and ϕ with the polar angle and the azimuthal angle of a spherical surface and thus arrive at the *Bloch sphere*, a useful visualization of the states of two-level systems.

The surface of the Bloch sphere describes the unit sphere in three-dimensional space by specifying the angles θ and ϕ (Fig. 3.18). If we consider the polarization of light as an example (with $|0\rangle = |H\rangle$ and $|1\rangle = |V\rangle$), we can determine the location of the states $|H\rangle$, $|V\rangle$, $|+\rangle$ and $|-\rangle$ by comparing with Eq. (3.72): The $|H\rangle$ state lies at the north pole, $|V\rangle$ at the south pole of the Bloch sphere; $|+\rangle$ and $|-\rangle$ lie at the equator with $\phi = 0$ and $\phi = \pi$, respectively. These four states lie on a circle described by real coefficients α and β. We have not mentioned so far the states with circular position. Their basis states $|R\rangle$ and $|L\rangle$ (for right and left circular polarization) contain complex coefficients:

$$|R\rangle = \frac{1}{\sqrt{2}}(|H\rangle + i|V\rangle), \quad |L\rangle = \frac{1}{\sqrt{2}}(|H\rangle - i|V\rangle). \tag{3.73}$$

π pulses and $\frac{\pi}{2}$ pulses   ℹ️

When qubits are realized by systems with two energy levels (e. g., ions or superconducting qubits), π-pulses or $\frac{\pi}{2}$-pulses are often mentioned when talking about gate operations. The meaning of these terms can be explained using the Bloch sphere.

Qubits are manipulated in these systems by interaction with the electromagnetic field, i. e., by laser or microwave pulses. The interaction induces transitions between the qubit states. Mathematically, the interaction changes the coefficients α and β in Eq. (3.71). The stronger the electric field of the pulse and the longer the pulse duration, the stronger the interaction.

Often, certain recurring standard operations on qubits are performed. The pulses required for this are designated accordingly:

1. A π-pulse rotates the state vector on the Bloch sphere by 180°. It converts the state $|0\rangle$ into $|1\rangle$ and vice versa.

2. A $\frac{\pi}{2}$ pulse rotates the state on the Bloch sphere by 90°. This creates superposition states. For example, the states $|+\rangle$ and $|-\rangle$ are produced from $|0\rangle$ and $|1\rangle$ by applying a $\frac{\pi}{2}$ pulse.

> ⚡ The Bloch sphere is a useful illustration of the states of two-level systems and, in particular, of the operations performed on them. However, it has one peculiarity that is confusing at first. *Orthogonal* States (like $|H\rangle$ and $|V\rangle$) are *opposite* on the sphere. So, the angle between them is 180° and not 90°. Mathematically, this is due to the appearance of the angle $\theta/2$ in Eq. (3.72), which is introduced in this way to ensure that the surface of the sphere is completely covered.

3.9 More complex quantum systems

So far, we have considered only the simplest quantum system: a single qubit with two possible states. We can extend this simplest model system in two directions: We can consider quantum objects with more states and systems composed of multiple quantum objects.

Quantum objects with several levels

If we increase the number of independent basis states of a quantum object, this is reflected mathematically by a higher dimensionality of the state vector $|\psi\rangle$. For example, if we consider a quantum object with three energy levels, then the state vector has three complex components, and so on. The scalar product for two n-dimensional state vectors $|a\rangle$ and $|b\rangle$ in an n-dimensional state space is:

$$\langle a|b\rangle = \sum_{i=1}^{n} a_i^* \, b_i. \tag{3.74}$$

Operators are described accordingly by $n \times n$ matrices.

If continuous variables occur, such as position or momentum, finite vectors are no longer sufficient to specify the state of the system. The state space becomes infinite-dimensional. The state vector is then represented by continuous *wave functions*, for example, in the form $\psi(x)$ as a function of position. Continuous variables and wave functions typically occur in traditional quantum mechanics, for example, in the description of electrons in atoms. In quantum technologies, finite-dimensional systems are more common. We will, therefore, not elaborate on the continuous case.

Multiple-qubit states

In the applications of quantum technologies, there are usually *multiple* quantum objects interacting with each other, for example, the qubits in a quantum computer. The execution of quantum algorithms requires multiple qubits, which are brought to interaction by gate operations. Let us first assume that we are dealing with *distinguishable* quantum objects, which can be characterized, e. g., by a well-defined position in a crystal lattice, on a chip, or in an ion trap.

As an example, we consider a two-qubit system. The two individual qubits are described by the states $|a\rangle$ and $|b\rangle$. The state of the total system can then be described by the *tensor product* of the two individual states:

$$|\psi\rangle = |a\rangle \otimes |b\rangle \quad \text{or} \quad |\psi\rangle = |a\rangle_1 |b\rangle_2. \tag{3.75}$$

We prefer the second notation, where the indices indicate to which qubit the single state refers. The scalar product for the two-qubit states can be decomposed pairwise into the scalar products of the single states. For the two states $|\psi\rangle = |a\rangle \otimes |b\rangle$ and $|\phi\rangle = |c\rangle \otimes |d\rangle$, it reads:

$$\langle\psi|\phi\rangle = \langle a|c\rangle \langle b|d\rangle, \tag{3.76}$$

or, in the notation with indices:

$$\langle\psi|\phi\rangle = ((\langle a|_1 \langle b|_2)(|c\rangle_1|d\rangle_2) = \langle a|c\rangle_1 \langle b|d\rangle_2. \tag{3.77}$$

The state (3.75), which is the product of two single-system states, is by no means the most general state of a two-qubit system. In general, if $|a_i\rangle$ and $|b_j\rangle$ are the basis states of the two systems, then any superposition is possible:

$$|\psi\rangle = \sum_{ij} c_{ij} |a_i\rangle_1 |b_j\rangle_2, \tag{3.78}$$

where the coefficients c_{ij} have to satisfy the normalization condition.

Example: Specify a basis for the state space of a two-qubit system.

Solution: We consider the basis states $|0\rangle_1$ and $|1\rangle_1$ for the first qubit and $|0\rangle_2$ and $|1\rangle_2$ for the second qubit. By forming tensor products, four normalized and mutually orthogonal states can be constructed from them:

$$|00\rangle = |0\rangle_1|0\rangle_2, \quad |01\rangle = |0\rangle_1|1\rangle_2,$$
$$|10\rangle = |1\rangle_1|0\rangle_2, \quad |11\rangle = |1\rangle_1|1\rangle_2. \tag{3.79}$$

As an example, we only show the orthogonality of $|01\rangle$ and $|10\rangle$:

$$\langle 01|10\rangle = \big((\langle 0|_1 \langle 1|_2)(|1\rangle_1|0\rangle_2)\big) = \langle 0|1\rangle_1 \langle 1|0\rangle_2 = 0. \tag{3.80}$$

As there are four orthogonal basis states, the state space for the two-qubit system is four-dimensional.

Example: Show that the state space of an n-qubit system is described by 2^n complex coefficients.

Solution: The state space of n qubits is spanned by the basis vectors $|00\ldots00\rangle$, $|00\ldots01\rangle$, $|00\ldots10\rangle$, $|00\ldots11\rangle$, ..., $|11\ldots11\rangle$ (each with n digits). The entries can be read as binary representations of the numbers from 0 to $2^n - 1$. Since the general state of the n qubit system is a superposition of all these states, 2^n complex coefficients are needed to describe it.

Thus, the dimensionality of the state space increases exponentially with the number of qubits. This is the main reason for the efficiency of quantum computers.

Matrix notation: While the tensor product can be represented quite naturally in the Dirac notation, it turns out to be more clumsy in matrix notation. We stay with the example of the two-qubit state. From two two-dimensional state vectors (a, b) and (c, d) we want to construct a four-dimensional one. This is done in the following way:

$$\begin{pmatrix} a \\ b \end{pmatrix} \otimes \begin{pmatrix} c \\ d \end{pmatrix} = \begin{pmatrix} a \times \begin{pmatrix} c \\ d \end{pmatrix} \\ b \times \begin{pmatrix} c \\ d \end{pmatrix} \end{pmatrix} = \begin{pmatrix} ac \\ ad \\ bc \\ bd \end{pmatrix}. \tag{3.81}$$

In this representation, the four basis vectors for a two-qubit system are:

$$|00\rangle = \begin{pmatrix} 1 \\ 0 \\ 0 \\ 0 \end{pmatrix}, \quad |01\rangle = \begin{pmatrix} 0 \\ 1 \\ 0 \\ 0 \end{pmatrix}, \quad |10\rangle = \begin{pmatrix} 0 \\ 0 \\ 1 \\ 0 \end{pmatrix}, \quad |11\rangle = \begin{pmatrix} 0 \\ 0 \\ 0 \\ 1 \end{pmatrix}. \tag{3.82}$$

Systems of indistinguishable quantum objects

So far, we have assumed that the quantum objects in the system under consideration are individually distinguishable by physical features such as a well-defined position. If this is not the case, different rules apply. For example, in a helium atom, there are two electrons that cannot be distinguished in any way. This is not just a practical problem (in the sense that we cannot tell them apart experimentally) but a new fundamental principle of quantum physics: Systems of multiple elementary particles of the same kind that are not made distinguishable by some attribute must be treated as indistinguishable quantum objects.

It depends on the type of particle and how this is represented in the formalism. There are *fermions*, whose spin is a half-integer multiple of the Planck constant $\hbar$, and *bosons* with integer spin. Electrons, protons, and neutrons are fermions, photons are bosons. The following rule applies when constructing the states of indistinguishable quantum objects: bosons have wave functions that remain unchanged when the two particles are formally exchanged; for fermions, the sign changes:

$$\psi(\vec{r}_1, \vec{r}_2) = \pm \psi(\vec{r}_2, \vec{r}_1). \tag{3.83}$$

The upper sign applies to bosons and the lower sign to fermions. The difference may not seem dramatic, but its consequences cannot be overestimated. For fermions, it implies the *Pauli principle* (two electrons can never be in exactly the same quantum state), which is fundamental to atomic physics. For bosons, conversely, the consequence is that they tend to "clump" in the same quantum state, leading to phenomena such as superconductivity, Bose-Einstein condensation, or the principle of the laser.

3.10 Decoherence and Schrödinger's cat

The superposition principle (3.6), which seems so inconspicuous at first sight, poses particularly large problems for the understanding of quantum physics. It states that if $|\psi_1\rangle$ and $|\psi_2\rangle$ are physically possible states of a quantum system, then all superposition states

$$|\psi\rangle = \alpha |\psi_1\rangle + \beta |\psi_2\rangle \tag{3.84}$$

are physically possible states of the system as well. The superposition principle is one of the fundamental postulates of quantum physics, and its validity is considered universal. In the double-slit experiment and the Mach-Zehnder interferometer, we have seen examples of how superposition leads to interference and how superposition states can be detected based on the occurrence of interference.

The cat paradox

If we accept the universal validity of the superposition principle, we run into serious difficulties. Why is it possible to observe superposition states of microscopic objects but never macroscopic bodies? The superposition principle should apply to them as well, and there should be superposition states of balls, stones, bicycles, and all kinds of other objects in our environment. Schrödinger had already pointed out this problem in 1935, and with his cat paradox, he had already invalidated one possible objection: It could be that the superposition states of macroscopic bodies exist in principle but are so difficult to prepare that they never succeed in practice. With the cat paradox, he gives a scheme for their production [29]:

> One can also construct quite burlesque cases. A cat is locked in a steel chamber together with the following infernal machine (which must be secured against direct access by the cat): In a Geiger counter tube there is a tiny amount of radioactive substance, so small that in the course of an hour perhaps one of the atoms will decay, but just as likely none will; when this happens the counter tube reacts and through a relay operaties a small hammer which smashes a small ball of hydrogen cyanide. If you let the whole system run for an hour, one would say that the cat is still alive if no atom has decayed in the meantime. The first atomic decay would have poisoned it. The ψ-function of the whole system would express this by having in it the living and the dead cat mixed or smeared in equal parts.

Schrödinger's scheme for generating superposition states for macroscopic bodies is shown in Fig. 3.19: A microscopic quantum object, for which we are able to generate superposition states, interacts with an amplification mechanism and thus transfers the superposition to a macroscopic object.

The fact that Schrödinger chose a living object for his thought experiment is spectacular, but it is rather an obstacle to understanding the paradox. In every discussion, the question of how to physically characterize the biological states "alive" and "dead" is raised quickly. It is very difficult to answer but also quite irrelevant to the question

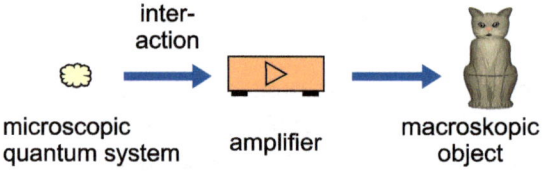

Figure 3.19: The cat paradox illustrates a scheme for generating macroscopic superposition states by an amplification process.

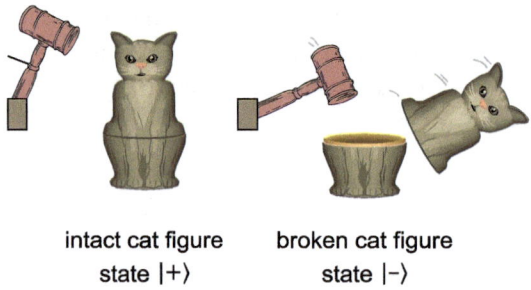

intact cat figure　　　broken cat figure
state $|+\rangle$　　　　state $|-\rangle$

Figure 3.20: Intact and broken cat figure as macroscopic object in Schrödinger's cat paradox.

of whether superposition states of macroscopic bodies exist or not. Just as well, we can consider, as in Fig. 3.20, a cat figure smashed by the hammer when the atom decays.

Superposition states and statistical mixtures

The question of how we would even recognize a superposition state of intact and broken cat figures is hardly ever raised in the debate about Schrödinger's cat. Are there measurements that could distinguish the superposition state from an ensemble of "classically intact" and "classically broken" figures?

We have already answered this question in connection with the double-slit experiment. On p. 43, we distinguished superposition states from "either-or" states (left slit or right slit). The experimental difference was the appearance of interference. The same applies here: The experimental evidence for superposition states is the appearance of interference.

ℹ With increasingly sophisticated techniques, it is possible to show interference of larger and larger objects. The interference experiment with whole helium atoms discussed above was an important breakthrough in 1991. In 1999, the interference of C_{60} molecules caused great excitement. The race goes on: in 2019, Arndt's group in Vienna succeeded in making large organic molecules of more than 2000 atoms interfere [30]. The mass of the molecules was about 25,000 atomic mass units. The progress shows: Superposition states of large objects are not impossible to produce. However, detecting them by the appearance of interference is an enormous experimental challenge.

In the mathematical formalism, quantum mechanical superposition states are distinguished from statistical mixtures. These are "either-or" states, classical ensembles in which each cat figure actually has the property "intact" or "broken". We consider the measurement of an observable D on an ensemble of cat figures. The task is to find out whether the ensemble under consideration is a classical statistical mixture or a quantum mechanical superposition state.

For a statistical mixture, the classical mathematical definition of the mean value of D is:

$$\langle D \rangle = \frac{N_+}{N} D_+ + \frac{N_-}{N} D_-, \tag{3.85}$$

where $\frac{N_+}{N}$ and $\frac{N_-}{N}$ are the relative frequencies of the intact and broken cat figures in the ensemble. D_+ and D_- are the values of the observable D in the respective state. For illustration, we can think of an example where D is the height of the cat's nose above the ground, although this is certainly not a realistic example.

For the quantum mechanical superposition state of the intact and the broken cat figure, $|\psi\rangle = \alpha \, |+\rangle + \beta \, |-\rangle$, the expectation value of D is calculated according to Eq. (3.60) with $\langle D \rangle = \langle \psi | D | \psi \rangle$. After expanding the terms, we get

$$\langle D \rangle = |\alpha|^2 \langle +|D|+ \rangle + |\beta|^2 \langle -|D|- \rangle$$
$$+ \alpha^* \beta \langle +|D|- \rangle + \alpha \beta^* \langle -|D|+ \rangle . \tag{3.86}$$

The terms in the first line structurally correspond to the classical formula (3.85); they can be interpreted as the classical mean value with the probabilities $|\alpha|^2$ and $|\beta|^2$ (i. e., relative frequencies). The deviation from the classical mean value are the cross terms in the second line, indicating the occurrence of interference. The experimental challenge is to choose the observable D so that, if a suitable parameter is varied, these interference terms can be detected.

Decoherence: the influence of the natural environment

The theory of decoherence explains why it is so difficult to observe quantum mechanical interference of macroscopic bodies. It is based on the observation that, in reality, macroscopic objects are never completely isolated. They always interact with their *natural environment*, which has a large number of degrees of freedom. This environment may be, for example, the ubiquitous thermal radiation whose photon distribution is determined by Planck's formula. Other examples are the interaction with the surrounding gas molecules or the ordinary daylight, which usually interacts strongly with macroscopic bodies (that is the reason why you can see them).

The interaction with the environment causes the object, taken by itself, to lose its quantum mechanical coherence over time because it builds up correlations with the environment. This process is called *decoherence* (cf. p. 47). One of the first to point out

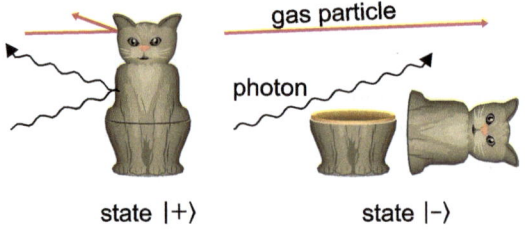

gas particle

photon

state $|+\rangle$ state $|-\rangle$

Figure 3.21: The broken cat figure affects the environment (photons and gas particles) in a different way than the intact one.

the importance of the natural environment was Zeh [31]; the theory was later developed further by Zurek [32] and others.

To explain the argument, we formally assign the state $|U\rangle$ to the environment of the cat figure – a state we neither know nor can control because it describes, in the order of 10^{23} gas molecules, as well as innumerable photons in the visible and infrared regions. The environment has very many degrees of freedom interacting with the cat figure; its state is the tensor product of all gas and photon states:

$$|U\rangle = |\text{gas}\rangle_1 |\text{gas}\rangle_2 \cdots |\text{gas}\rangle_{10^{23}} |\text{photon}\rangle_1 \cdots |\text{photon}\rangle_N. \tag{3.87}$$

Crucial to the argument is that the broken cat figure affects the environment in a *different* way than the intact one (Fig. 3.21): It scatters light in a different way, gas molecules collide with surfaces in different places and are deflected differently. Thus, the broken cat figure will be correlated with the environment differently than the intact one. This will already be the case after an extremely short interaction time. The state of the total system of cat figure and environment takes the following form:

$$|\psi\rangle = \alpha |+\rangle |U_+\rangle + \beta |-\rangle |U_-\rangle. \tag{3.88}$$

The correlation between the cat figure and the environment is expressed by the fact that different environment states belong to $|+\rangle$ and to $|-\rangle$. The state (3.88) can no longer be written as a product in the form (state of the cat) × (state of the environment). Cat and environment are *entangled* – a term coined by Schrödinger in this context, which has gained enormous importance in modern quantum physics (cf. Section 3.14).

Again, we consider measurements of the cat figure observable D, which acts only on the states of the cat figure but not on the states of the environment. Analogous to Eq. (3.86), we obtain:

$$\langle D \rangle = |\alpha|^2 \langle +|D|+\rangle \underbrace{\langle U_+|U_+\rangle}_{=1} + |\beta|^2 \langle -|D|-\rangle \underbrace{\langle U_-|U_-\rangle}_{=1}$$
$$+ \alpha^* \beta \langle +|D|-\rangle \underbrace{\langle U_+|U_-\rangle}_{\approx 0} + \alpha \beta^* \langle -|D|+\rangle \underbrace{\langle U_-|U_+\rangle}_{\approx 0}. \tag{3.89}$$

The scalar products of the environment states in the first line have the value 1 because the vectors are normalized that way. More interesting are the scalar products in the second line, which are zero to an excellent approximation so that in total:

$$\langle D \rangle \approx |\alpha|^2 \, \langle +|D|+\rangle + |\beta|^2 \, \langle -|D|-\rangle . \tag{3.90}$$

The interference terms vanish because we can assume that the environment states $|U_+\rangle$ and $|U_-\rangle$ are orthogonal. The resulting equation (3.90) has the same form as the formula for the classical mean value (3.85), which occurs in the statistical mixture of broken and intact cat figures. This non-appearance of interference due to the interaction of a macroscopic body with its environment is called decoherence.

Why is it justified to assume that the environment states $|U_+\rangle$ and $|U_-\rangle$ are orthogonal? Their scalar product is a product of very many terms (cf. Eq. (3.77)), all of which have modulus ≤ 1. If only one of the factors is zero, then the entire product is zero. This happens, for example, if just one of the gas molecules or photons is scattered in a completely different way by the intact and the broken figure. Even if this is not the case: If each of the factors is only slightly smaller than 1, the product becomes effectively zero because of the sheer number of factors. To illustrate the order of magnitude: If we have a scalar product of 10^{23} factors, and each of them has the value 0.999, we arrive at $(0.999)^{10^{23}} \approx e^{-10^{20}}$, an incredibly small number. The mechanism of decoherence is so effective that the time scales on which the ability to interfere is destroyed are extraordinarily short. Estimates with simple models give decoherence times of typically 10^{-30} s.

> The theory of *decoherence* describes how macroscopic objects lose their ability to interfere due to uncontrollable interactions with their environment. They thus become "effectively classical".

Schrödinger's question of why there are no superposition states of macroscopic bodies is thus answered by the theory of decoherence: Macroscopic bodies appear classical because they cannot be isolated from their environment. Interaction with the environment destroys their ability to interfere. The cat figure is intact or broken; superposition or interference phenomena cannot be detected.

This is bad news for the construction of quantum computers. The working principle of quantum computers is essentially based on superposition states and interference between different qubits. A working quantum computer with 50 qubits is thus in a macroscopic quantum state similar to the case of Schrödinger's cat. Even worse, its state has to be completely controlled in every detail. To suppress decoherence as much as possible, only systems that can be well isolated from their environment are considered for the realization of quantum computers.

Vacuum and low temperatures help to shield against gas molecules and infrared photons. Particularly suitable are systems that are naturally "immune" to certain external influences, e. g., ions against infrared photons.

We have seen that, in principle, the scattering of one single gas molecule or photon into orthogonal states is sufficient to prevent interference through the mechanism of decoherence. This is the physical background of the criterion for the occurrence of interference that we formulated in connection with the principle of complementarity (basic rule 4). On p. 46, it was expressed as follows:

„To prevent interference, it is sufficient for the quantum objects to leave a trace somewhere in the environment, from which one could in principle infer which of the classical alternatives has been realized."

This is a qualitative description of how interference is suppressed by decoherence. In this version, it turns from a statement about physical interaction processes into an information-theoretical criterion: The ability to interfere is lost when information about the state of the system escapes into the environment. This makes the basic rule a powerful and easy-to-use tool in qualitative argumentation.

3.11 The density matrix formalism

Normally the state of a quantum mechanical system is described by the state vector $|\psi\rangle$. It contains all the information about the system that can be obtained in accordance with the indeterminacy relations. To obtain this information, measurements on a complete set of commuting observables (whose eigenstates completely span the state space) or a suitable preparation have to be made.

Completely and incompletely known systems

How are systems described if this maximal information about the system is not available? For example, one could have completely characterized the spatial degrees of freedom of an ensemble of photons but not the polarization. It is then impossible to say which of the polarization states $|H\rangle$ or $|V\rangle$ (or superpositions of them) should be attributed to the ensemble. There is a subjective ignorance of the experimental situation, i. e. not all possibilities of preparation compatible with the indeterminacy relations were used. In short, there are *uncontrolled degrees of freedom.*

The occurrence of subjective ignorance or uncontrolled degrees of freedom is the characteristic of a *statistical mixture* – a term we have already used on p. 81 to distinguish it from *pure states*. Pure states are characterized by complete information and the use of all preparation possibilities. They can be represented by state vectors. For the description of statistical mixtures, the formalism has to be generalized: Uncontrolled degrees of freedom and subjective ignorance are described in quantum mechanics by the *density matrix* formalism.

The density matrix

In the example of photon polarization, we can only give probabilities for the states $|H\rangle$ and $|V\rangle$: p_H for $|H\rangle$ and p_V for $|V\rangle$. Without any knowledge about the polarization, p_H and p_V are equal, so $p_H = p_V = \frac{1}{2}$. This is called *unpolarized light.*

To calculate the mean value of any observable A for a statistical mixture, one proceeds according to the rules of classical probability theory (as in Eq. (3.85) or Eq. (3.90)) and forms the sum of the (quantum mechanical) individual mean values weighted by the probabilities p_H and p_V:

$$\langle A \rangle_{\text{mixture}} = p_H \langle H|A|H \rangle + p_V \langle V|A|V \rangle . \tag{3.91}$$

An elegant description for this procedure is obtained by introducing the *density matrix ρ*:

$$\rho = \begin{pmatrix} p_H & 0 \\ 0 & p_V \end{pmatrix} = p_H |H\rangle\langle H| + p_V |V\rangle\langle V|. \tag{3.92}$$

With the density matrix, the mean value (3.91) is calculated by multiplying the two matrices ρ and A and forming the trace Tr of the product matrix ρA (which is done by adding the diagonal elements):

$$\langle A \rangle = \text{Tr}(\rho A) = (\rho A)_{HH} + (\rho A)_{VV}. \tag{3.93}$$

Interpretation of the density matrix

To understand the density matrix it is crucial to recognize the two different concepts of probability used here: on the one hand the interpretation of probability inherent in quantum mechanics which allows only probabilistic statements about the results of measurements. It is a fundamental feature of quantum mechanics and cannot be overcome by obtaining additional information. On the other hand, there is our subjective ignorance about uncontrolled degrees of freedom, which could be eliminated by additional measurements.

The difference can be best illustrated by comparing the density matrix for a statistical mixture of the two states $|H\rangle$ and $|V\rangle$,

$$\rho = \begin{pmatrix} p_H & 0 \\ 0 & p_V \end{pmatrix}, \tag{3.94}$$

with the one of the pure state $|\psi\rangle = \alpha |H\rangle + \beta |V\rangle$, which is a coherent superposition of $|H\rangle$ and $|V\rangle$. The density matrix for a pure state $|\psi\rangle$ is generally given by

$$\rho' = |\psi\rangle\langle\psi|. \tag{3.95}$$

After inserting $|\psi\rangle = \alpha |H\rangle + \beta |V\rangle$, we get:

$$\rho' = \begin{pmatrix} |\alpha|^2 & \alpha\beta^* \\ \alpha^*\beta & |\beta|^2 \end{pmatrix}. \tag{3.96}$$

We note that in the density matrix (3.96) of the pure state there are off-diagonal elements, which are zero in Eq. (3.94). It is these off-diagonal elements that are responsible for the interference between $|H\rangle$ and $|V\rangle$. This can be seen by calculating the mean value (3.93) of an observable A for the two density matrices (3.94) and (3.96):

$$\langle A \rangle_{\text{mixture}} = \text{Tr}(\rho A) = p_H \langle H|A|H \rangle + p_V \langle V|A|V \rangle \,,$$

$$\langle A \rangle_{\text{pure}} = \text{Tr}(\rho' A) = |\alpha|^2 \langle H|A|H \rangle + |\beta|^2 \langle V|A|V \rangle + \alpha^* \beta \langle H|A|V \rangle + \alpha\beta^* \langle V|A|H \rangle \,.$$

The two expressions differ just by the interference terms between $|H\rangle$ and $|V\rangle$, which we already know from Eq. (3.86). Mathematically, they result from the off-diagonal elements of the density matrix (3.96).

⚡ Unpolarized light

The density matrix for unpolarized light expressed in terms of the H/V basis is given by the normalized unit matrix (Eq. (3.94) with ($p_H = p_V = \frac{1}{2}$)). The question arises: If we do not have any information about the polarization state of the light, why can we say that the state is uniformly mixed from the states $|H\rangle$ and $|V\rangle$ and not from the alternative basis states $|+\rangle$ and $|-\rangle$?

The answer is that we do not know this at all. In fact, the density matrix in the +/− basis (and any other basis) is exactly the same as in the H/V basis: Regardless of the basis used, the density matrix for unpolarized light is the normalized unit matrix. This is welcome for the logical consistency of the formalism. However, it also means that unpolarized light incoherently mixed from H/V states cannot be distinguished experimentally in any way from unpolarized light incoherently mixed from +/− states – a puzzling result that invites further reflection.

Quantum mechanical description of a subsystem

A strength of the density matrix formalism is that it allows the description of *subsystems*. This is often needed in the field of quantum information. We consider a coupled system consisting of two subsystems 1 and 2. We assume that the total system is described by a pure state. According to Eq. (3.78), its state vector can be composed of the tensor products of the subsystem basis vectors:

$$|\psi\rangle = \sum_{ij} c_{ij} |a_i\rangle_1 |b_j\rangle_2. \tag{3.97}$$

If the two subsystems interact, it is generally not possible to assign a separate state vector to each of them. The correlations established by the interaction between 1 and 2 imply that the two subsystems are not independent of each other. Thus, the state of system 1 depends on the state of 2 and vice versa, as in Eq. (3.88). In this case, the two systems are *entangled*.

While it is not possible to specify a state vector to describe subsystem 1 alone, one can introduce a quantity that does something similar: the *reduced density matrix* ρ_{red}, which is obtained from the total state (3.97) by tracing over the degrees of freedom of subsystem 2:

$$\rho_{\text{red}} = \text{Tr}_2(|\psi\rangle \langle\psi|) = \sum_k {}_2\langle b_k|\psi\rangle \langle\psi|b_k\rangle_2 \,. \tag{3.98}$$

The sum extends over a complete system of states of subsystem 2. Because of the trace over subsystem 2, ρ_{red} acts solely in the state space of subsystem 1. Here, the reduced density matrix behaves as expected: If we restrict ourselves to observables A_1 which refer only to subsystem 1, ρ_{red} allows the calculation of all expectation values as if only subsystem 1 were present, described the density matrix ρ_{red} – that is, by forming the trace over subsystem 1 in Eq. (3.93):

$$\langle A_1 \rangle = \text{Tr}_1(\rho_{red}A_1). \tag{3.99}$$

One thus obtains the same results as with the full state vector (3.97) and forms the trace over both subsystems: $\langle A_1 \rangle = \text{Tr}_{1,2}(|\psi\rangle \langle\psi| A_1)$.

The formal differences expressed in the theoretical description of a quantum object by a pure state $|\psi\rangle$, on the one hand, and the reduced density matrix ρ_{red}, on the other hand, have a relevant physical background. The pure state describes a *closed system* which can evolve undisturbed by external influences. In contrast, the reduced density matrix refers to an *open subsystem* that is part of a larger system and is constantly interacting with its environment.

Description of decoherence with the density matrix formalism

The density matrix formalism is the appropriate way to describe decoherence because the quantum system under consideration is only a subsystem of a larger system that also includes the environment, whose many degrees of freedom we do not know and cannot control. We had written in Eq. (3.88) for the state of the total system of quantum system plus environment:

$$|\psi\rangle = a\,|+\rangle\,|U_+\rangle + \beta\,|-\rangle\,|U_-\rangle\,. \tag{3.100}$$

In the density matrix formalism, we assign a reduced density matrix to the quantum system taken by itself. It is obtained from the state (3.100) by tracing over the environment variables:

$$\rho_{red} = \text{Tr}_{Umgebung}\big(|\psi\rangle \langle\psi|\big) = \sum_i \langle U_i|\psi\rangle \langle\psi|U_i\rangle\,. \tag{3.101}$$

The sum extends over a complete system of environment states $|U_i\rangle$. For all measurements performed only on the quantum system, the reduced density matrix yields the same values as the complete state (3.100).

The formation of the trace in Eq. (3.101) corresponds to an averaging over the uncontrollable and unobservable degrees of freedom of the environment which influence the system. If we evaluate the reduced density matrix (3.101) for the state (3.100) and use the fact that, according to the previous reasoning, the environmental states belonging to the object states $|+\rangle$ and $|-\rangle$ are orthogonal to a very good approximation ($\langle U_+|U_-\rangle \approx 0$), we find:

$$\rho_{red} \approx |a|^2\,|+\rangle\,\langle+| + |\beta|^2\,|-\rangle\,\langle-| = \begin{pmatrix} |a|^2 & 0 \\ 0 & |\beta|^2 \end{pmatrix}. \tag{3.102}$$

The reduced density matrix is diagonal in the $+/-$-basis. From the disappearance of the off-diagonal elements, we see that the occurrence of interference is effectively suppressed. The density matrix is the same as (3.92) with the classical probabilities $p_+ = |a|^2$ and $p_- = |\beta|^2$.

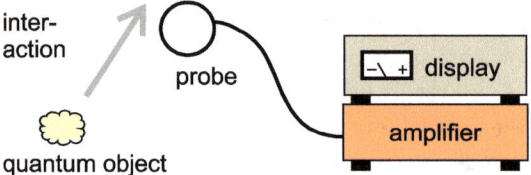

Figure 3.22: Schematic representation of a quantum mechanical measurement process.

3.12 The quantum mechanical measurement process

In the debate about the interpretation of quantum physics, the measurement process has been discussed more intensively than any other topic. The books and papers on this subject fill many shelves. The advent of decoherence theory has changed the situation. Since its widespread acceptance in the late 1990s, the measurement problem can be considered largely settled, at least for all practical purposes. However, since there are still many inaccurate or outdated accounts, it will be addressed here.

Basically, it is possible to describe measurements in quantum mechanics by the two rules stated above: The Born probability formula and the requirement that the possible outcomes of a measurement are the eigenvalues of the measured observable. This is a phenomenological description that is sufficient in many practical cases.

The measurement problem

However, we can go one step further and describe the process of measurement itself in quantum mechanical terms. Thus, in the theory of measurement, the measuring device itself is included in the quantum mechanical description. Since quantum mechanics is claimed to be universally valid, it should also be able to describe the process of measurement.

Historically, this has led to a problem known as the "quantum mechanical measurement problem". The problem can be described as follows: During a measurement, a microscopic quantum object interacts with a macroscopic apparatus via a probe (Fig. 3.22). The interaction should change the state of the apparatus so that the desired information about the object can be read from the apparatus (e. g. by noting the pointer position). This is the minimum requirement for a reasonable measurement.

It turns out that the situation is equivalent to Schrödinger's cat paradox. As shown in Fig. 3.19, the state of the microscopic quantum object is linked to that of a macroscopic object (the apparatus) by a process of amplification. If the states of the microscopic object are denoted by $|+\rangle$ and $|-\rangle$ and the states of the apparatus by $|M_+\rangle$ and $|M_-\rangle$, then the system consisting of the object and measuring device is in a superposition state after the interaction:

$$|\psi\rangle = \alpha\,|+\rangle\,|M_+\rangle + \beta\,|-\rangle\,|M_-\rangle\,. \tag{3.103}$$

As it should be, the measured object and measuring device are correlated: The values + or − are linked to the corresponding pointer position of the measuring device. In this sense, our experiment has achieved the desired goal.

You would expect from a measurement that you could read the result from the state of the device: The pointer should be either at the + position or at the − position. However, this is not the case in the state (3.103). As in the case of Schrödinger's cat, the apparatus does not end up in a state $|M_+\rangle$ or $|M_-\rangle$ with a well-defined pointer position but in a superposition of both. Operationally, the superposition could be detected by interference between the two components of the state (3.103) – although it is not easy to imagine how such an interference experiment would actually be performed.

Decoherence in the measurement process

As in the case of Schrödinger's cat, the problem is solved by including the environment. According to the decoherence theory, the description of a measurement by the state (3.103) is too idealized. A more realistic description of the measurement process would include the environment and the decoherence process. Analogous to Eq. (3.88), the state of the system including the environment is:

$$|\psi\rangle = \alpha\,|+\rangle\,|M_+\rangle\,|U_+\rangle + \beta\,|-\rangle\,|M_-\rangle\,|U_-\rangle\,. \tag{3.104}$$

We trace over the uncontrolled environment degrees of freedom, and, as in Eq. (3.102), we assume that $\langle U_+|U_-\rangle \approx 0$. This gives us the reduced density matrix:

$$\rho_{\mathrm{red}} = |\alpha|^2\,|+, M_+\rangle\,\langle +, M_+| + |\beta|^2\,|-, M_-\rangle\,\langle -, M_-| = \begin{pmatrix} |\alpha|^2 & 0 \\ 0 & |\beta|^2 \end{pmatrix}. \tag{3.105}$$

Its off-diagonal elements are zero; all interference is destroyed by decoherence. The reduced density matrix (3.105) describes a statistical mixture, an ensemble of "effectively classical" measurement devices which are in "either-or states" with a well-defined pointer position + or −.

State reduction

What is the state of the measured object after the measurement? There is no general answer to this question. For example, a photon is absorbed by the detector during detection. It is destroyed by the measurement. It is different from a helium atom in the double-slit experiment of p. 39. Although the atom is delocalized over the whole detection region immediately before the measurement, it is found at a specific position on the screen. Afterwards, it is still detectable there. A second position measurement will give the same result as the first.

This particular type of measurement, where the second measurement confirms the result of the first measurement, is quite common. It is described in the formalism as follows: After the first measurement, we take note of its result and thus obtain new information. With this new information, we update our description of the system. We assign a new density matrix to the system in which only the measured value is represented. The density matrix "collapses" to a single entry. This is basically the same process as lifting a dice cup and noting the result. It is a process that does not take place in the real world but only affects our description of it.

Usually, the process is described in less detail, and the following is stated: If in a measurement of the observable A the measured value a_i was found, then after the measurement, the system is in the corresponding eigenstate $|a_i\rangle$. This process is called *state reduction* or *collapse of wave function*. Once you are aware of the underlying assumptions and processes, this is an accurate shorthand description in many cases.

The above description of state reduction assumes that the state vector or density matrix is theoretical entity that describes reality but has no physical reality itself. State reduction is then conceived as an "update" of our description of the system as we acquire new information about it. There is no epistemological problem in this interpretation.

Often, however, the opinion is held that the state vector has an independent physical reality, i. e. that it really exists "out there". Then the state reduction must seem puzzling. How can it be that after a measurement, only one component of the state remains, and all others disappear? A causal mechanism for this cannot be given, and the measurement problem remains a serious problem of quantum mechanics. The most radical way out of this dilemma is Everett's "many-worlds theory", which is readily taken up in the popular science literature. The speculation about "disappearing" state components is related to the question of where all that nice money went after a stock market crash. The question only makes sense under the assumption that this money actually existed before, and then it is very difficult to answer.

For practical application, the statistical interpretation described in the previous sections, including decoherence, is the most appropriate. It avoids epistemological problems that have no experimental or practical implications anyway (no experiment has ever failed because the theory of the quantum mechanical measurement process was not properly understood). The arguments for the statistical interpretation of quantum physics are lucidly and convincingly presented by Englert in his article *On Quantum Theory* [33].

Finally, we summarize what we can be said in general about measurements in quantum mechanics:

Rules of quantum mechanical measurement:
1. The possible outcomes of a measurement are the eigenvalues of the measured observable.
2. Generally, the results of individual measurements cannot be predicted (except in cases where the probability is 0 or 1). If the measurement is repeated many times on an ensemble of identically prepared quantum objects, probability statements are possible. The probabilities can be determined by Born's probability formula (3.13).
3. For some types of measurements, the state after the measurement is the eigenstate of the measured observable that belongs to the measured value.

3.13 Bell's inequality

The inequality that John Bell found in 1964 is not easy to understand at first glance. This is due to its epistemological status, which needs some explanation. Even though it is considered to be one of the most important statements about quantum mechanics – there's not the slightest mention of quantum mechanics in it. Bell's inequality is a statement about *classical alternative theories* to quantum mechanics. Its formulation was motivated by a widespread uneasiness about the probabilistic character of quantum mechanics and the departure from classical determinism. This discomfort was succinctly expressed by Albert Einstein in a letter to Max Born in 1926 [34]:

> The theory produces a good deal but hardly brings us closer to the secret of the Old One. I am at all events convinced that He does not play dice.

Here Einstein raises the possibility that certain "peculiar" features of quantum mechanics (probabilistic character, non-attribution of properties) may not reflect nature itself but merely point to a deficiency in the theory. There was hope for a more complete alternative theory with "hidden variables" that are not captured by quantum mechanics and that deterministically specify all measurement results in advance – in particular, even for those variables that, according to quantum mechanics, cannot be assigned fixed values at the same time (such as position and momentum or the polarization components of light).

Bell considers these "local-realistic alternative theories" to quantum mechanics (one of which, Bohmian mechanics, had already been formulated at that time). However, he did not investigate any particular one of these alternative theories. Bell's inequality is a statement about *all* local-realistic alternative theories that satisfy certain conditions. It is quite remarkable that such a general statement is possible at all, and it took a long time for the scientific community to properly understand the implications of Bell's inequality. At the time of its publication, it did not seem particularly impressive. Reinhold Bertlmann, a close associate of John Bell, recalls [35]:

> At CERN, John was a kind of oracle for particle physics, consulted by many colleagues who wanted to get his approval for their ideas. Of course, I had heard that he was also a leading figure in quantum mechanics – specifically, in quantum foundations. But nobody, either at CERN or anywhere else, could actually explain his foundational work to me. The standard answer was, „He discovered some relation whose consequence was that quantum mechanics turned out all right. But we knew that anyway, so don't worry."

Local-realistic alternative theories to quantum mechanics

Bell's inequality requires very few and plausible assumptions about the alternative theories under consideration. They should satisfy two requirements (from which the designation "local-realistic" derives):

1. *Realism:* Quantum systems are completely described by a list of parameters (hidden variables) that are fixed at the time of preparation and determine in advance the results of all measurements of the system's observables. In short, all observables have fixed values before the measurement.

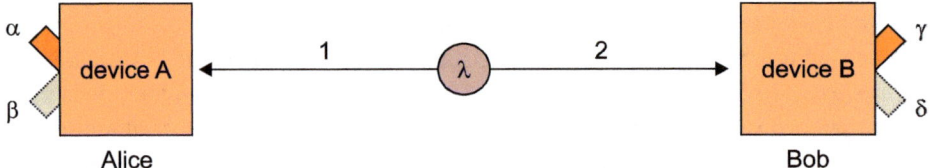

Figure 3.23: Schematic representation of an experiment for testing Bell's inequality. Two quantum objects are emitted from a common source and propagate to spatially separated measuring devices A and B. The observable to be measured can be selected independently on each of the two devices.

2. *Locality:* If the system consists of two spatially separated subsystems A and B, then the results of measurements in A should not depend on which observables are measured in B.

To derive Bell's inequality, we consider the simplest possible case (Fig. 3.23): Two quantum objects 1 and 2, which are prepared together but do not interact afterwards, are sent to two spatially separated observers (Alice and Bob) with measurement devices A and B, respectively. The measurement devices can optionally measure one of two observables α and β (on Alice's side) or γ and δ (on Bob's side). For each new quantum object arriving, Alice and Bob can decide, independently of each other, which of the two observables to measure (indicated by the two switch positions in Fig. 3.23). For simplicity, we assume that the observables have only two possible values, which are scaled to ±1.

Relevant conclusions arise if, in quantum mechanics, an indeterminacy relation holds for the two observables under consideration, i. e. if they are complementary (most of the experiments that have been performed to test the Bell inequalities used the different polarizations of light). In the following argument, however, there is no reference to quantum mechanics. We have already mentioned that Bell's inequality is not a statement about quantum mechanics but about possible local-realistic alternative theories with hidden parameters.

Formulation of Bell's inequality
We consider an alternative theory in which the properties of the pair emitted in Fig. 3.23 are described by a set of hidden parameters λ of some kind. They deterministically fix all future measurement results in advance. Thus, all observables have fixed values independent of any measurement; they are ordinary numbers, which we denote in the calculations by A_α, A_β, B_γ, and B_δ. The notation $A_\alpha = +1$ indicates that Alice measured the variable α and found the value to be +1; while $B_\delta = -1$ means that Bob measured the variable δ and found it to be –1. Now we examine the following expression:

$$A_\alpha\,(B_\gamma + B_\delta) + A_\beta\,(B_\gamma - B_\delta). \tag{3.106}$$

Since all of the variables can only take the values +1 or −1, two cases can be distinguished for each measurement: either $(B_\gamma + B_\delta) = 0$, in which case $(B_\gamma - B_\delta) = \pm 2$; or $(B_\gamma - B_\delta) = 0$, then $(B_\gamma + B_\delta) = \pm 2$. In any case, the expression (3.106) is equal to ± 2, so its absolute value is ≤ 2. This is still true when taking the expectation value over many measurements. In this way, we obtain the *Bell-CHSH inequality*:

$$\left| \langle A_\alpha B_\gamma \rangle + \langle A_\alpha B_\delta \rangle + \langle A_\beta B_\gamma \rangle - \langle A_\beta B_\delta \rangle \right| \leq 2. \tag{3.107}$$

This variant of Bell's inequality is due to Clauser, Horne, Shimony, and Holt (CHSH) [36]. It holds for all variables $A_\alpha, A_\beta, B_\gamma, B_\delta$ that can take the values ± 1 and have a joint probability distribution determined by the parameters λ.

Experiments to test Bell's inequality have been successfully performed since the 1970s, mainly with the polarization degrees of freedom of entangled photons (Physics Nobel Prize 2022 for Alain Aspect and John Clauser). Today, they can be carried out with the highest precision and avoid various "loopholes". The experiments show that there are certain states for which Eq. (3.107) is *not* satisfied. Thus, the Bell-CHSH inequality does not hold in general; it has been disproved by experiment. At least one of the assumptions used to derive it must therefore be false. If we want to remain in agreement with the experiment, we are forced to drop either the assumption of locality or the classical realistic description.

The understanding of the Bell-CHSH inequality is complicated by the fact that it is difficult to see which of the assumptions that led to Eq. (3.107) could possibly be wrong. Everything seems to follow in the simplest way, from elementary mathematics and pure logic. The point is subtle: it is the assumption that the observables A_α and A_β or B_γ and B_δ have fixed numerical values at the same time, i. e. that they can appear together as ordinary numbers in an equation like Eq. (3.106). So, the flaw in the argument is that we have simultaneously assigned definite values to two complementary observables.

In quantum mechanics, it is mathematically impossible to formulate a relation like Eq. (3.106), which holds for ordinary numbers. It is only the expectation value of the product of the operators in Eq. (3.107), which is well defined in quantum mechanics. This expectation value can be compared with the experiment. As we will discuss later, the experiments brilliantly confirm the prediction of quantum mechanics.

Significance of Bell's inequality

In quantum mechanics, the first of the two assumptions above is abandoned: the classical realistic description in which all observables have definite values, whether they are measured or not. According to quantum mechanics, quantum objects often cannot even be assigned certain properties, and even less are the outcomes of a measurement fixed in advance. Measurements that are not performed have no results in quantum mechanics.

A viable alternative would be to abandon the assumption locality (this possibility is realized, for example, in Bohmian mechanics). However, locality is a central requirement of Einstein's theory of relativity. It is crucial for relativity that causal influences cannot propagate faster than the speed of light. Most physicists consider the compatibility with relativity too important to go this route.

The experimentally established violation of Bell's inequality, together with the rejection of explicit nonlocality, thus leads to the abandonment of classical determinism. It also provides the justification for basic rule 1 (statistical behavior): If there are no hidden variables that determine the result of a measurement in advance, then the result is indeed random. Thus, the violation of Bell's inequality implies that measurement results are not predetermined; there is *objective randomness* in nature. This randomness is not based on subjective ignorance but is inherent in the phenomena.

⚡ You often hear about the nonlocality of quantum mechanics. How is this compatible with what has been said above? We have just heard that it is precisely the locality which is not to be abandoned.

Here again is a subtle point: It is not quantum mechanics itself that is nonlocal in the sense of the second criterion of p. 92, but those classical, alternative theories (like Bohmian mechanics) that are able to successfully describe the experimental results.

In this sense, quantum mechanics is local because it satisfies the aforementioned criterion: In no way can Alice's experimental results or probabilities be affected by anything Bob does. He cannot send messages to Alice by selecting the variables he measures (i. e. by choosing switch position 1 or 2 in Fig. 3.23). The "perceived" nonlocality of quantum mechanics is due to the fact that the entangled states used in the experiments are spread over a region of space that includes both observers. The states are not localized, and measurements on them show strong correlations. Any "mechanistic" model that would allow us to visualize the correlations would be a special case of a hidden variable theory and therefore would have to be either nonlocal or wrong (most often both are the case).

ℹ️ **The Kochen-Specker theorem**

The *Kochen-Specker theorem* [37] is logically independent of Bell's inequality but aims in the same direction. Unlike Bell's inequality (which assumes pairs of quantum objects), it is already applicable to single quantum objects. It deals with the problem of hidden parameters from another perspective. It examines the question of whether *non-contextual* assignments of definite values to observables are possible. That is, is it possible to assign certain properties to an object independently of what other variables are measured?

We consider measurements on two observables, X and Y, of a quantum object. The two measurements are assumed to be compatible. This is the case if they can be performed in any order, successively or simultaneously, without disturbing the results. Whether two measurements are compatible can be verified experimentally. In quantum mechanics, this is the case if $[X, Y] = 0$, i. e. if the observables commute.

The Kochen-Specker theorem refers to three observables A, X, Y. It is assumed that $[A, X] = [A, Y] = 0$ and $[X, Y] \neq 0$. Thus, the observable A is compatible with both X and Y, but X is not compatible with Y. The Kochen-Specker theorem states (by explicitly giving a counterexample) that, in general, it is not possible to assign a value to the observable A that is independent of whether it is measured together with X or with Y. This is particularly surprising because both X and Y are compatible with A. Therefore, it is possible to measure them only after the measurement of A has long been done. It is not conceivable how such a scenario can be realized if A represents an "element of reality" which has an existence independent of a specific measurement situation.

Regarding possible alternative theories to quantum mechanics, the Kochen-Specker theorem states: All theories in which the observable A is assigned a definite value, regardless of whether it is measured together with X or Y, lead to predictions other than those of quantum mechanics and thus contradict experiment.

3.14 Entanglement

The term entanglement, coined by Schrödinger in 1935 in the same paper in which he introduced the cat paradox, denotes one of the most important and characteristic differences between classical and quantum physics. It occurs in the context of quantum systems composed of two or more subsystems. These can be the decaying atom and the cat of Fig. 3.19 or the two quantum objects, which, in the context of Bell's inequality, are sent to Alice and Bob (Fig. 3.23).

In general, two quantum objects 1 and 2 are *unentangled* or *separable* if their state vector can be written as the product of the two individual states:

$$|\psi\rangle = |\psi_1\rangle \otimes |\psi_2\rangle. \tag{3.108}$$

Here, $|\psi_1\rangle$ refers only to quantum object 1 and $|\psi_2\rangle$ refers only to quantum object 2. If this is not the case, then the two quantum objects are *entangled*. In this case, they can then no longer be described as individual objects, but form a system whose overall state cannot be factorized.

> Two quantum objects are entangled if the total state of the system cannot be decomposed into a product of individual states.

Example: Consider a system of two qubits. Show that the state

$$|\Psi^-\rangle = \frac{1}{\sqrt{2}}\left(|+-\rangle - |-+\rangle\right) \tag{3.109}$$

is entangled.

Solution: In general, states of single qubits have the form of superpositions:

$$|\psi_1\rangle = a|+\rangle + b|-\rangle \quad \text{and} \quad |\psi_2\rangle = c|+\rangle + d|-\rangle. \tag{3.110}$$

The general product state is the tensor product of these two states:

$$|\psi_1\rangle \otimes |\psi_2\rangle = \left(a|+\rangle + b|-\rangle\right) \otimes \left(c|+\rangle + d|-\rangle\right)$$
$$= ac|++\rangle + ad|+-\rangle + bc|-+\rangle + bd|--\rangle. \tag{3.111}$$

Every unentangled state of the total system can be written in this form. If this is not possible for the state under consideration, it is entangled. By comparing the coefficients with Eq. (3.109), we obtain the following four conditions:

$$ac \overset{!}{=} 0, \quad ad \overset{!}{=} \frac{1}{\sqrt{2}}, \quad bc \overset{!}{=} -\frac{1}{\sqrt{2}}, \quad bd \overset{!}{=} 0. \tag{3.112}$$

From the first condition, it follows that at least one of the factors a and c must be zero. For $a = 0$, however, the second condition can no longer be fulfilled, and similarly, $c = 0$ contradicts the third condition. Thus, there are no coefficients a, b, c, d that satisfy all four conditions. A factorization is not possible; the state (3.109) is entangled.

Characteristics of entangled states

We have already encountered entangled states in the discussion of the measurement problem and the cat paradox (Eq. (3.88) and (3.103)). The arguments there show that entangled states are nothing unusual in quantum mechanics. They arise whenever two quantum objects interact, provided that the interaction depends on their internal state (or other local degrees of freedom). In this sense, entangled states are ubiquitous.

If entanglement is nothing unusual in quantum mechanics, why did it take so long to prove it experimentally? This was only achieved in the 1980s in the experiments of Aspect et al. on Bell's inequality [38]. The answer to this question was already hinted at in the discussion of Schrödinger's cat paradox: it is not the generation of entanglement per se that poses experimental difficulties, but the *controlled* generation of entanglement and, above all, its detection.

---

i **Creation of entangled photon pairs**

For photons, the controlled generation of entangled polarization states is possible today with relatively little effort by *spontaneous parametric down-conversion* (parametric fluorescence) in BBO crystals (cf. p. 15). This method for generating polarization-entangled photon pairs was introduced by Kwiat et al. in 1995 [39]. In parametric fluorescence, an incident photon is converted into two outgoing photons in a nonlinear crystal. Due to energy conservation, each of the outgoing photons has half the energy of the incoming photon, i. e., twice the wavelength.

In the experiment, a pump laser (e. g., a high-energy diode laser with a wavelength of 405 nm in the ultraviolet range) is focused on the BBO crystal. With a very small probability, two photons with a wavelength of 810 nm are produced, which, due to the conservation of momentum, are emitted on an emission cone (each at the same angle to the optical axis and in the same plane as the pump laser).

To generate entangled photon pairs, two identical BBO crystals are used whose anisotropy axes are perpendicular to each other. Depending on the orientation, vertically polarized photons incident on this double crystal can only be converted into horizontally polarized photon pairs in the first crystal. Conversely, horizontally polarized light can only be converted into vertically polarized pairs in the second crystal. However, if light with linear polarization is incident at an angle of 45° to the two anisotropy axes of the crystals, conversion is possible in both crystals with the respective polarizations. The result is a superposition of the two possibilities:

$$|\Phi\rangle = \frac{1}{\sqrt{2}}\left(|HH\rangle + e^{i\phi}|VV\rangle\right). \tag{3.113}$$

The relative phase ϕ depends, among other things, on the thickness of the crystals and can be controlled, for example, by the relative phase between the horizontal and vertical components of the pump laser. Alternatively, a superposition of the emission cones of the two BBO crystals is achieved with the help of additional optical components. The photon pairs of both types (horizontally polarized and vertically polarized) thus become indistinguishable with respect to the BBO crystal in which they originated. Thus, the two maximally entangled Bell states (which we will discuss in more detail below)

$$|\Phi^{\pm}\rangle = \frac{1}{\sqrt{2}}\left(|HH\rangle \pm |VV\rangle\right) \tag{3.114}$$

can be created. The other two Bell states

$$|\Psi^{\pm}\rangle = \frac{1}{\sqrt{2}}\left(|HV\rangle \pm |VH\rangle\right) \tag{3.115}$$

can also be produced via additional optical, birefringent components. The pairs of photons created in this way are entangled not only with respect to their polarization but also with respect to their energy and momentum.

The experimental demonstration of entanglement is done by evaluating correlations. Alice and Bob perform their respective measurements in the experimental scheme of Fig. 3.23 and then compare their results. Entanglement is revealed by the appearance of *correlations* that are stronger than a classical description would allow. These strong correlations are detected by the violation of Bell's inequality (or a variant of it). This is – besides the non-existence of hidden variables and the existence of objective randomness – another reading of Bell's inequality.

Correlations are the experimental hallmark of entanglement.

We have already seen examples for correlations in entangled states in Eq. (3.88) and (3.103): the dead cat belongs to the decayed atom, and the object state $|+\rangle$ belongs to the pointer position M_+ of the measurement apparatus. Taken by themselves, however, such correlations are not yet specific for quantum mechanics because they also occur in classical physics. For example, if Alice has a bag with a red and a blue ball in it, and Bob draws a ball without looking, then there is a correlation between the color of the ball in Bob's hand and the color of the ball in Alice's bag, without this being remarkable in any way.

Correlations that are relevant to quantum mechanics occur when two observables are considered that are complementary in quantum mechanics, i. e. that cannot be assigned definite values at the same time. The state (3.109) – often used in experiments and called the *singlet state* – shows the correlations for complementary observables particularly clearly.

Correlations in the singlet state
We consider the singlet state (3.109):

$$|\Psi^-\rangle = \frac{1}{\sqrt{2}}(|+-\rangle - |-+\rangle). \tag{3.116}$$

It can be physically realized with a variety of two-state systems, such as the +/– polarization states of light. If Alice and Bob both measure the polarization of their photon in the +/– basis, as in Fig. 3.24, the (anti-)correlation can be read directly from Eq. (3.116): Whenever Alice finds the value +, Bob obtains – and vice versa. They will never find the same value. Expressed more formally using Born's probability formula:

$$P(+-) = \left|\langle + - |\Psi^-\rangle\right|^2 = \frac{1}{2}; \quad P(-+) = \left|\langle - + |\Psi^-\rangle\right|^2 = \frac{1}{2}, \tag{3.117}$$

and $P(++) = P(--) = 0$. As described earlier, these correlations are not remarkable by themselves. It becomes interesting only if we add a second pair of observables and

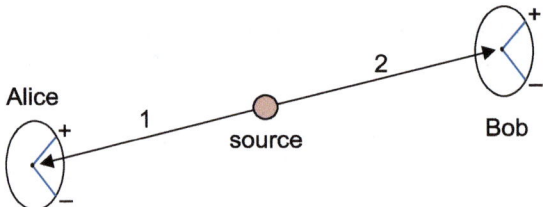

Figure 3.24: Schematic setup for measuring polarization correlations.

make measurements, e. g., in the H/V basis. The correlations that appear then can be determined most easily if we transform the state (3.116) into the H/V basis. It turns out that in this basis, it has the *same* form as above:

$$|\Psi^-\rangle = \frac{1}{\sqrt{2}}(|HV\rangle - |VH\rangle). \tag{3.118}$$

Thus, for H/V measurements, the same anticorrelations between Alice's and Bob's results occur as for +/− measurements. Considering the result from p. 71 that H/V polarization and +/− polarization cannot be prepared without scatter at the same time because an indeterminacy relation holds for them and that therefore definite values cannot be assigned to both properties at the same time, this result seems most surprising.

However, that is not all. As we will show in the following example, the singlet state has the same form in *all* orthogonal polarization bases. No matter what measurement basis Alice and Bob agree on, they will always find the same complete anticorrelation in their results. With this characteristic, the singlet state violates Bell's inequality. Many experiments on Bell's inequality are based on such states.

Example: Show that the singlet state is invariant under a general unitary transformation of the basis.

Solution: We write the unitary transformation that mathematically describes the change to the new basis $|y^+\rangle$, $|y^-\rangle$ in the following general form,

$$|y^+\rangle = a\,|+\rangle + b\,|-\rangle, \quad |y^-\rangle = -b^*\,|+\rangle + a^*\,|-\rangle, \tag{3.119}$$

with complex coefficients a and b where $|a|^2 + |b|^2 = 1$. A special case are the rotations with $a = \cos y$ and $b = \sin y$ (cf. Eq. (3.64); the different sign in this equation means only an irrelevant global phase in one of the states). The inverse transformation equations are:

$$|+\rangle = a^*\,|y^+\rangle - b\,|y^-\rangle, \quad |-\rangle = b^*\,|y^+\rangle + a\,|y^-\rangle. \tag{3.120}$$

We now substitute these expressions into Eq. (3.116), which in more detailed notation reads:

$$|\Psi^-\rangle = \frac{1}{\sqrt{2}}\left[|+\rangle_1|-\rangle_2 - |-\rangle_1|+\rangle_2\right]. \tag{3.121}$$

We obtain:

$$|\Psi^-\rangle = \frac{1}{\sqrt{2}}\left[\left(a^*|\gamma^+\rangle_1 - b|\gamma^-\rangle_1\right)\left(b^*|\gamma^+\rangle_2 + a|\gamma^-\rangle_2\right)\right.$$
$$\left. - \left(b^*|\gamma^+\rangle_1 + a|\gamma^-\rangle\right)\left(a^*|\gamma^+\rangle_2 - b|\gamma^-\rangle_2\right)\right]. \tag{3.122}$$

After expanding and taking advantage of $|a|^2 + |b|^2 = 1$, we get:

$$|\Psi^-\rangle = \frac{1}{\sqrt{2}}\left[|\gamma^+\rangle_1|\gamma^-\rangle_2 - |\gamma^-\rangle_1|\gamma^+\rangle_2\right]. \tag{3.123}$$

The singlet state has the same form also in the new basis. It is thus invariant under the basis transformation (3.119); it has the same form in every basis. Physically, this means the complete anticorrelation of Alice's and Bob's measurements in the experiment of Fig. 3.22, no matter at which angle their polarization measurement is performed (as long as both choose the same angle).

If Alice and Bob do not choose equal settings for their measuring devices but measure at different angles, they will get less strong correlations. The probabilities can be calculated using Born's probability formula and the transformation equations (3.119):

$$P(+,\gamma^+) = \frac{1}{2}|b|^2, \quad P(+,\gamma^-) = \frac{1}{2}|a|^2, \quad P(-,\gamma^+) = \frac{1}{2}|a|^2, \quad P(-,\gamma^-) = \frac{1}{2}|b|^2.$$

Maximally entangled states

Alice and Bob get the perfect correlations described above only if they perform their measurements on the same basis and then compare their results. For this comparison, they must meet or send physical signals. Both can be done at most at the speed of light so that the relativistic causality is preserved despite the quantum mechanical correlations. In general, transferring information from one place to another by measurement is not possible in quantum mechanics. In the case of the singlet state, Alice and Bob will get completely random results when they perform their measurements individually, without comparing them – regardless of the basis on which they measure. Mathematically, this is reflected in the fact that the reduced density matrix resulting from tracing over one of the subsystems is proportional to the unit matrix. States for which this is the case are called *maximally entangled*.

Example: Calculate the reduced density matrix that results for Alice when the trace over Bob's subsystem is taken for the singlet state.

Solution: We have learned about the description of a subsystem by the reduced density matrix on p. 86. To obtain the reduced density matrix for Alice, we consider the density matrix of the singlet state $\rho = |\Psi^-\rangle\langle\Psi^-|$:

$$\rho = \frac{1}{2}\left[|+\rangle_1|-\rangle_2 - |-\rangle_1|+\rangle_2\right]\left[\langle+|_1\langle-|_2 - \langle-|_1\langle+|_2\right]. \tag{3.124}$$

We take the trace over Bob's subsystem (index 2):

$$\rho_{\text{red}} = \text{Tr}_2\left(|\Psi^-\rangle\langle\Psi^-|\right)$$
$$= \sum_{i=\pm} \frac{1}{2}\langle i|_2\left[|+\rangle_1|-\rangle_2 - |-\rangle_1|+\rangle_2\right]\left[\langle+|_1\langle-|_2 - \langle-|_1\langle+|_2\right]|i\rangle_2. \tag{3.125}$$

Only two of the four terms resulting from the multiplication are different from zero, and the result is:

$$\rho_{\text{red}} = \frac{1}{2}\left[\underbrace{\langle+|+\rangle}_{=1} {}_2|-\rangle_1 \langle-|_1 \underbrace{\langle+|+\rangle}_{=1} {}_2 + \underbrace{\langle-|-\rangle}_{=1} {}_2|+\rangle_1 \langle+|_1 \underbrace{\langle-|-\rangle}_{=1} {}_2 \right]$$

$$= \frac{1}{2}\left[|-\rangle \langle-|_1 + |+\rangle \langle+|_1 \right] = \frac{1}{2}\begin{pmatrix} 1 & 0 \\ 0 & 1 \end{pmatrix}. \tag{3.126}$$

Thus, the reduced density matrix for Alice is diagonal, and the probabilities for the outcomes $+$ or $-$ are both $\frac{1}{2}$. This is true not only for the $+/-$ basis but also for any other basis. Thus, Alice will get completely random results from her local measurements, from which she cannot extract any information. This is true no matter what Bob does (e. g., what settings he makes on his device). Alice cannot discover the correlations contained in the state $|\Psi^-\rangle$ by local measurements. This can only be done in retrospect by comparison with Bob's results.

If only Alice's local measurements are considered, the description by the reduced density matrix is equally appropriate as the description by the pure state $|\Psi^-\rangle$. All probabilities for Alice's local measurements are correctly described. This is an example of how the state vector or the density matrix is not a unique feature of the quantum system itself. They only represent an observer's description of the system. Different observers, depending on their information about the system, may have different descriptions of it, all of which can be equally valid [33].

The state $|\Psi^-\rangle$ is one of four states that together form the so-called *Bell basis*:

Bell basis: The four states of the Bell basis form a complete orthonormal system for describing two-qubit states:

$$|\Phi^+\rangle = \frac{1}{\sqrt{2}}\left(|++\rangle + |--\rangle\right), \quad |\Phi^-\rangle = \frac{1}{\sqrt{2}}\left(|++\rangle - |--\rangle\right),$$

$$|\Psi^+\rangle = \frac{1}{\sqrt{2}}\left(|+-\rangle + |-+\rangle\right), \quad |\Psi^-\rangle = \frac{1}{\sqrt{2}}\left(|+-\rangle - |-+\rangle\right). \tag{3.127}$$

All four states of the Bell basis are maximally entangled. However, only the singlet state is invariant under the general unitary transformation (3.119). For the state $|\Phi^+\rangle$, this is the case if a and b are real, i. e., in particular for the rotations of polarization analyzers described above. Besides the standard two-qubit basis, $|00\rangle$, $|01\rangle$, $|10\rangle$ and $|11\rangle$, the Bell basis is one of the most commonly used bases to describe two-qubit states.

Entanglement entropy

In addition to maximally entangled states, such as the Bell basis, there are also weakly entangled states. Using the reduced density matrix, a measure of the degree of entanglement can be defined for pure states, the *entanglement entropy*:

$$S(\rho_{\text{red}}) = -\text{Tr}_1(\rho_{\text{red}} \log_2(\rho_{\text{red}})). \tag{3.128}$$

Here ρ_{red} is the density matrix traced over subsystem 2. Writing the expression in a basis where ρ_{red} is diagonal gives:

$$S(\rho_{\text{red}}) = -\sum_i \lambda_i \log_2(\lambda_i), \tag{3.129}$$

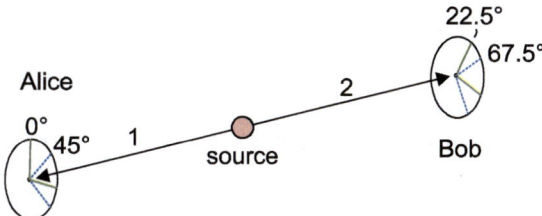

Figure 3.25: Angle settings for testing the Bell-CHSH inequality.

where the λ_i are the eigenvalues in this basis. The expression (3.129) has the same form as the *Shannon entropy* in classical information theory. The entanglement entropy is zero for non-entangled (separable) states while it is 1 for the maximally entangled states (3.127):

$$S(\rho_{\mathrm{red}}) = S\left(\frac{1}{2}\mathbb{1}\right) = -\frac{1}{2}\log_2\frac{1}{2} - \frac{1}{2}\log_2\frac{1}{2} = 1. \tag{3.130}$$

For more weakly entangled pure states, the value of the entanglement entropy is between 0 and 1. For mixed states, it is more difficult to find appropriate entanglement measures; there are different approaches here.

Violation of Bell's inequality in the singlet state

In Section 3.13, the Bell-CHSH inequality (3.107) was formulated, and it was claimed that there are quantum states for which it is violated. The singlet state $|\Psi^-\rangle$ is one of them. To demonstrate the violation of the Bell-CHSH inequality, we need to calculate correlations of the form $\langle A_\alpha B_\gamma\rangle$. To clarify the mathematical structure, we write the expression in more detail:

$$\langle\Psi^-|A_\alpha \otimes B_\gamma|\Psi^-\rangle. \tag{3.131}$$

A_α is an operator acting only on the states in subsystem 1 (Alice); the operator B_γ acts only on the states in subsystem 2 (Bob). They describe different settings of the respective measuring devices, e. g. polarization measurements at angles α for Alice and γ for Bob (Fig. 3.25). The term (3.131) describes the expectation value in the state $|\Psi^-\rangle$. Perfect correlation should yield +1, perfect anticorrelation −1, and the uncorrelated case 0.

Example: Calculate the expectation value (3.131) for the singlet state $|\Psi^-\rangle$.

Solution: For the calculation, we exploit the symmetry of the singlet state. Since it has the same form in each basis, a common rotation of the two polarization analyzers in Fig. 3.25 by the same angle cannot affect the measurement results. Consequently, the result can only depend on the angular difference $\gamma - \alpha$. Therefore, to simplify the calculation, we let Alice's analyzer point in the +/− direction; γ is the angle with respect to this direction. We substitute $|\Psi^-\rangle$ into Eq. (3.131) and expand the terms:

$$\langle A_a B_y \rangle = \frac{1}{2} \big[\langle +|A_a|+\rangle_1 \langle -|B_y|-\rangle_2 - \langle -|A_a|+\rangle_1 \langle +|B_y|-\rangle_2$$
$$- \langle +|A_a|-\rangle_1 \langle -|B_y|+\rangle_2 + \langle -|A_a|-\rangle_1 \langle +|B_y|+\rangle_2 \big].$$

(3.132)

Since Alice's device is oriented in the $+/-$ direction, $|+\rangle_1$ and $|-\rangle_1$ are eigenstates of A_a with eigenvalues $+1$ and -1. If we exploit this, the second and third terms vanish, leaving:

$$\langle A_a B_y \rangle = \frac{1}{2} \big[\langle -|B_y|-\rangle_2 - \langle +|B_y|+\rangle_2 \big].$$

(3.133)

We insert the transformation formulas (3.120) with $a = \cos y$ and $b = \sin y$ and use that these are the eigenstates of B_y. We obtain:

$$\langle A_a B_y \rangle = \frac{1}{2} \big[\big(|b|^2 - |a|^2\big) - \big(|a|^2 - |b|^2\big) \big] = \sin^2 y - \cos^2 y.$$

(3.134)

Using the mathematical identity $\sin^2 y - \cos^2 y = -\cos(2y)$, and because y represents the angular difference $y - a$ between Bob's and Alice's analyzers, the result is:

$$\langle A_a B_y \rangle = -\cos\big(2(y - a)\big).$$

(3.135)

For an angular difference of $0°$, perfect anticorrelation results, as expected for the singlet state. For an angular difference of $90°$, the roles of the $+$ and $-$ outputs are reversed for one of the observers, and this "relabeling" results in perfect correlation, in accordance with Eq. (3.135).

With the result (3.135), we can now check the violation of the Bell-CHSH inequality

$$\big| \langle A_a B_y \rangle + \langle A_a B_\delta \rangle + \langle A_\beta B_y \rangle - \langle A_\beta B_\delta \rangle \big| \le 2.$$

(3.136)

We use Eq. (3.135) to calculate the correlation functions, replacing a and y with other angles as needed. It turns out that the inequality is not violated for all angular positions. The maximum violation occurs when the angle settings satisfy the following conditions: $(y - \beta) = (a - y) = (\delta - a) = 22.5°$ and $(\delta - \beta) = 67.5°$. This is fulfilled, for example, for the following angles:

$$\text{Alice:} \quad a = 45°, \quad \beta = 0°,$$
$$\text{Bob:} \quad y = 22.5°, \quad \delta = 67.5°.$$

This results in:

$$\big| \langle A_a B_y \rangle + \langle A_a B_\delta \rangle + \langle A_\beta B_y \rangle - \langle A_\beta B_\delta \rangle \big| = 3\cos(45°) - \cos(135°) = 2\sqrt{2}.$$

With a value of $2\sqrt{2} \approx 2.83$ the Bell-CHSH inequality for the singlet state is clearly violated. Experimentally, the prediction of quantum mechanics is excellently confirmed. Nowadays, the violation of the Bell-CHSH inequality by many standard deviations is a routine experiment that belongs to the adjustment of experimental set-ups.

4 Quantum sensors

Of all the areas of quantum technologies, *quantum sensing* is currently the most advanced in terms of application maturity. In particular, the atomic clocks, discussed in Section 1.3, have long been used in technical applications ranging from high-precision time measurement to gravimetric measurements. *Metrology*, the science of measurement, is an important application of quantum sensing – for example, in the definition of SI units.

Much of quantum sensing is based on effects in single atoms, ions, molecules, or solids that can today be controlled so precisely that they allow measurement of a wide range of external influences. The unprecedentedly precise control of the systems allows the traditional logic of measurement to be reversed: Whereas uncontrollable environmental variables, such as magnetic fields or pressure, that affect the states of the system are normally sources of noise that have to be suppressed, the control of individual quantum systems has now advanced to the point where they can serve as sensors for these once-perturbing environmental variables. The potential applications of atomic clocks have already been discussed in Section 1.3. As a second quantum system with broad sensor applications; we will discuss NV centers, a special type of point defects in diamond crystals.

4.1 Measuring with NV centers

Natural crystals contain a wide variety of defects, deviations from the regularity of the crystal structure. Some of them form localized structures in which electrons can be bound, like in an artificial atom. These artificial atoms, with their well-defined quantized energy levels, can be used as sensors by exciting them with light and microwaves and detecting fluorescent light. This works in air and at room temperature, eliminating the need for vacuum and cryogenics – a major advantage over other systems.

NV centers in diamond

NV centers in diamond have proven to be particularly advantageous [40]. A diamond crystal is a regular arrangement of carbon atoms. Over 500 different crystal defects are known, such as single impurity atoms that disrupt the regular structure, or vacancies. An NV center consists of a nitrogen atom (N) and a vacancy (V) in the crystal lattice, which together can be thought of as a quantum mechanical system. This becomes clear when the electron structure of the defect is considered. The lattice structure for an NV center is shown schematically in Fig. 4.1.

A carbon atom has four valence electrons, which combine with the electrons of four other carbon atoms in the diamond crystal to form electron-pair bonds. This creates a very stable lattice structure. Nitrogen atoms have five valence electrons. Three of them

https://doi.org/10.1515/9783110717457-004

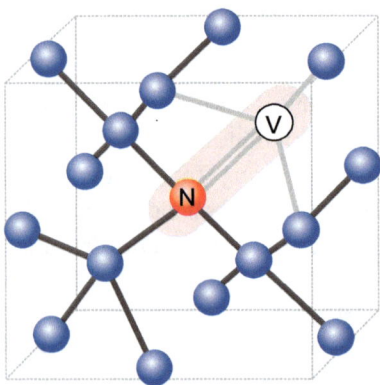

Figure 4.1: Lattice structure of an NV center in diamond with a nitrogen atom (N) and vacancy (V).

form electron-pair bonds with the surrounding carbon atoms in an NV center. The defect thus has one electron pair from the nitrogen atom and three electrons from the surrounding carbon atoms, making a total of five electrons. The defect is therefore uncharged and is called an NV^0 center. To obtain a manipulable quantum system, one more electron is needed, which can come from other impurities in the environment. This type of defect is relevant for use as a sensor and is called NV^- center or simply NV center.

ℹ Production of NV centers

Synthetic diamond can be produced in two main ways: either by high pressure and high temperature (HPHT) or by chemical vapor deposition (CVD). Also, synthetic diamond already contains nitrogen impurities and thus NV centers. Higher concentrations of NV centers are obtained by bombardment with high-energy electrons or ions and subsequent annealing at temperatures above 800 °C. Afterwards, the distances between the centers are still large enough to be studied individually.

NV centers as magnetic field sensors

As a localized structure, the NV center forms discrete energy levels. The levels relevant for quantum sensing are shown in Fig. 4.2. The transition between the spin triplet states $|g\rangle$ and $|e\rangle$ is in the optical range and can be excited by lasers (typically a diode laser at 532 nm); the emitted fluorescence light is detected optically. The use of NV centers as magnetic field sensors is based on the shift of energy levels in the magnetic field, the *Zeeman effect* discussed on p. 30.

The sublevels of the ground state $|g\rangle$ are crucial for the measurement. Without the magnetic field, the splitting between the states with $m = 0$ and $m = \pm 1$ is at 2.87 GHz. With increasing magnetic field, the energies of the states with $m = \pm 1$ split linearly due to the Zeeman effect (right in Fig. 4.2). The corresponding transitions are in the range of a few GHz and can be excited by microwave irradiation.

In the method called *Optically Detected Magnetic Resonance* (ODMR), the NV center is excited both optically and with microwaves. Bright fluorescent light from the opti-

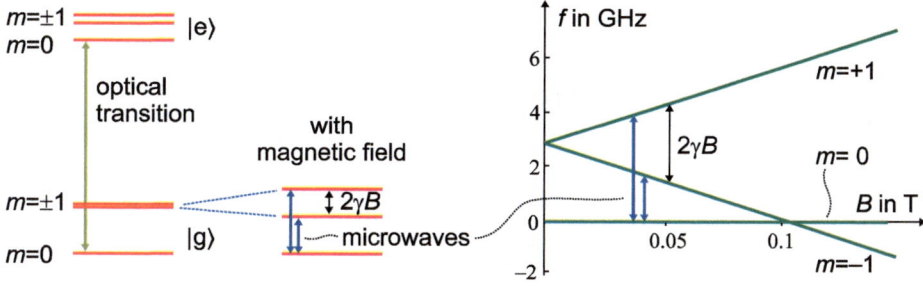

Figure 4.2: Left: Relevant energy levels when using NV centers as magnetic field sensors (not scaled). Right: Splitting of transitions as a function of the magnetic field.

cal excitation is observed when the laser light is in resonance with the optical transition shown in green in Fig. 4.2. At the same time, the microwaves induce the transitions marked with blue arrows. The microwave frequency is continuously swept to find the transition frequencies between the states with $m = 0$ and $m = \pm 1$. The resonance is indicated by a decrease in optical fluorescence because although the laser light can excite transitions from the states with $m = \pm 1$, the subsequent de-excitation is very likely to occur without the emission of visible light (it passes through intermediate states, not shown in the figure).

The magnetic field is measured by sweeping the microwave frequencies and registering the fluorescence rate. Due to the two resonances between $m = 0$ and $m = +1$ or $m = -1$, there are two minima in the fluorescence rate, whose distance is given by $\Delta f = 2\gamma B_\parallel$ with $\gamma = 28\,\text{MHz/mT}$ (Fig. 4.3). $B_\parallel$ is the component of the magnetic field in the direction of the NV center axis (shown in pink in Fig. 4.1). Thus, by measuring the distance between the two minima, the strength of the magnetic field component can be determined. In this way, the NV center can be used as a magnetic field sensor.

Other quantities such as electric fields, temperature, or pressure can also induce a splitting of the energy levels. In this way, NV centers can be used to measure many other quantities [41]. **i**

NV centers as qubits   **i**

In addition to their use as sensors, NV centers are also an approach for the physical realization of qubits, i. e., quantum computing hardware. The ground state levels from Fig. 4.2 with $m = 0$ and $m = -1$ are used as realizations of the qubit states $|0\rangle$ and $|1\rangle$. They can be read out by their different optical properties – similar to those described above for their use as sensors. Qubit operations (gates) can then be realized, for example, by microwave pulses that induce controlled transitions between levels. For example, a simple variant of the Grover algorithm (Section 6.3) has been realized using an NV center [42]. An advantage of NV centers is their long coherence time because they are well shielded against external influences (and thus decoherence) in the diamond crystal. A disadvantage for use as qubits is that their fabrication technique is hardly suitable for scaling to higher qubit numbers.

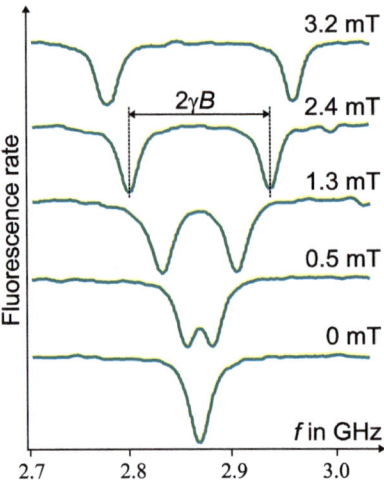

Figure 4.3: The magnetic field can be determined from the distance of the minima in the fluorescence spectrum (Data: QZabre).

4.2 Interaction-free quantum measurement

In addition to quantum sensor technologies, which, like atomic clocks or NV sensors, have evolved from traditional applications of quantum physics by achieving ever greater accuracy and control of the systems, there are approaches to measurement techniques that use quantum physical principles in novel ways. In 1993, a thought experiment by Elitzur and Vaidman [43], which came to be known as "interaction-free quantum measurement" caused quite a stir. Here, under certain circumstances, an object is detected without any physical interaction ever taking place with it. This seems to contradict our basic ideas of measurement, which normally cannot take place without interaction between the object and a probe.

The quantum bomb test – a thought experiment

Elitzur and Vaidman wrapped their measurement scheme in a thought experiment that is instructive but not very realistic: the *Quantum Bomb Test*. It involves testing hypothetical bombs whose detonators are triggered by a single photon. In the thought experiment, functional bombs with detonators are to be distinguished from those without detonators, without the bomb exploding. Classically, the problem seems insoluble: The interaction with at least one photon necessary for measurement would cause the bomb to explode.

The experimental setup considered in the thought experiment is a single photon Mach-Zehnder interferometer (Fig. 4.4; cf. p. 51). The bomb is placed in one arm of the interferometer. If it has a detonator, it protrudes into the path of the photons. A bomb without a detonator has no effect on the photons. The bomb itself is a purely classical

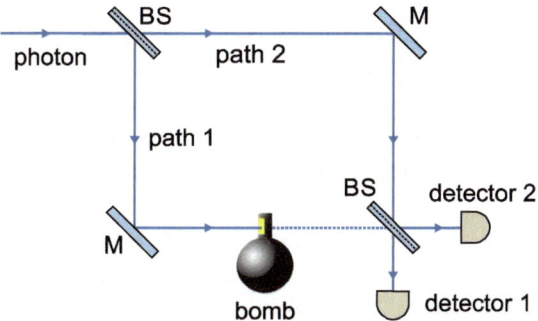

Figure 4.4: Quantum bomb test setup: Mach-Zehnder interferometer consisting of two beam splitters (BS), two mirrors (M), and two detectors (D_1 and D_2) into which a sensitive bomb is inserted.

object (the details of which are irrelevant to the thought experiment). All that matters is that it is made to explode by the absorption of a single photon.

Before the bomb is inserted, the path lengths in the interferometer are adjusted so that constructive interference occurs at detector D_1, and destructive interference occurs at detector D_2. Thus, without the bomb, all photons are detected in detector D_1 – at least in an idealized view, where we assume efficient and noise-free detectors. Detector D_2 never detects anything.

Now the bomb is placed in the interferometer. Every photon that reaches the detonator will be absorbed. When this happens, the bomb explodes. Three cases can occur, which are closely related to our discussion of the basic rules using the Mach-Zehnder interferometer as an example in section 3.3:

1. The photon is absorbed by the detonator of the bomb, and the bomb explodes. This happens in half of the cases (just as detector 1 in Fig. 3.8 clicks in half of the cases). In this case, nothing is gained. The experiment has to be repeated with a new bomb.
2. The photon is not absorbed, so it must have passed the interferometer arm without the bomb, i.e., path 2. At the second beam splitter, it can then be transmitted and detected in detector D_1 ...
3. ...or reflected and detected in detector D_2. Both occur with a probability of 25 %.

In the last case, i.e., in 25 % of the trials, we can conclude from the response of detector D_2 that a bomb with a detonator is in the beam path – without it having exploded. The interference between the two paths has been altered by the presence of the bomb: With the interruption of path 1, there are no longer two alternatives that are superposed at the second beam splitter, and no destructive interference occurs at detector D_2. From the absence of destructive interference, we can deduce the presence of the bomb. Only one photon was used in the whole experiment and this photon was detected in the detector. There is no photon that could have interacted with the detonator. Hence the name *interaction-free quantum measurement*.

In principle, what happens in this measurement scheme is that which-path information is obtained by absorption or non-absorption of the photon. If the detonator of the bomb does not absorb the photon, we can be sure that it went the other path – without direct interaction with the detonator. The measurement principle is essentially based on the alteration of the interference pattern when the which-path information is available (basic rule 4; cf. p. 46).

By no means all photons are actually detected at detector D_2 in the presence of a detonator. It is just as likely that the detection will occur at detector D_1 (case 2). Then we are not able to tell whether there is a bomb or not. The measurement must be repeated until one of the other two cases occurs.

 Example: Determine the probability of identifying a bomb with a detonator.

Solution: After the first beamsplitter, there is a probability of $\frac{1}{2}$ for the detection of the photon in the corresponding arm, i. e., also for the explosion of the bomb (case 1). For the other half of the photons, the same occurs again at the second beamsplitter so that the detection probability for both detectors is $\frac{1}{2} \times \frac{1}{2} = \frac{1}{4}$ (cases 2 and 3). Since case 2 leads to the repetition of the measurement, only the other two cases contribute to the overall result. Thus, the probability of successfully detecting a bomb with a detonator is

$$\frac{P(\text{case 3})}{P(\text{case 1}) + P(\text{case 3})} = \frac{\frac{1}{4}}{\frac{1}{2} + \frac{1}{4}} = \frac{1}{3}. \tag{4.1}$$

Increasing the probability of success
With a success probability of $\frac{1}{3}$, it is twice as likely that the bomb will explode during the measurement than that it will be detected as functional. To improve this ratio, an asymmetric beamsplitter can be used instead of a 50:50 beamsplitter. This may reduce the probability for case 1, while increasing the probability for cases 2 and 3. Thus, fewer bombs are detonated during the measurement. It turns out that the probability of success can thus be increased from $\frac{1}{3}$ to about $\frac{1}{2}$ [43]. With more complex schemes, a further increase of the detection probability up to almost 1 is possible [44].

4.3 Quantum imaging

The principle of interaction-free quantum measurement has inspired the development of several imaging techniques, collectively known as *Quantum imaging* [45, 46]. Pairs of entangled photons are used, one of which interacts with the object without contributing to an image, while the other the other photon is spatially resolved by a camera. In this way, an object can be imaged without the spatially resolved photons that are detected interacting directly with the object. This allows imaging at wavelengths where efficient detectors are unavailable.

Quantum Ghost Imaging: Imaging with correlated photons
For *Quantum Ghost Imaging*, the spatial and temporal correlations of an entangled photon pair are exploited (Fig. 4.5). The two photons are emitted from the nonlinear crystal

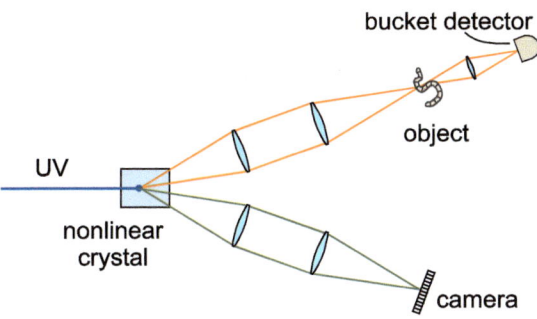

Figure 4.5: Principle of quantum ghost imaging with correlated photons.

in which they were generated. One of them interacts with the object. If it is not absorbed, it is – depending on the experimental configuration – reflected or transmitted and then detected by a non-resolving detector (in Fig. 4.5 above). The name "bucket detector" describes its function: like a bucket, it collects only the binary information whether a photon comes from the object or not. An image is not created.

The lower arm also does not produce an image of the object although the photons are detected there by a spatially resolving camera. However, since they have never interacted with the object, no image of the object is formed in the camera. It is just the light source that is imaged, i. e., the spatial profile of the pair generation processes in the nonlinear crystal.

An image of the object is formed only when the correlations between the photons are taken into account. To do this, one proceeds as follows: The position of a photon arriving at the spatially resolving camera is recorded only if a photon coming from the object is simultaneously registered at the bucket detector. In a sense, the photons that arrive at the bucket detector illuminate the object. The position information of the corresponding correlated partner photon is recorded by the camera. By coincidence detection, the image can be constructed in this way [47].

Quantum ghost imaging with different wavelengths

One potential application of quantum ghost imaging is efficient imaging in the infrared range. Typically, thermal imaging cameras are used for imaging in this region, where the detection of infrared radiation is based on arrays of microbolometers – thermal sensors that rely on heating. This operating principle results in low spatial resolution and low signal-to-noise ratio.

One possible application of quantum ghost imaging is to improve imaging in the infrared. For this purpose, a variant is used in which photons from different wavelength ranges are utilized for the two tasks (interaction with the object and spatially resolved detection). In the process of photon pair generation, an infrared photon is correlated with an optical photon. The infrared photon interacts with the object and provides its transmission or reflection properties in the infrared wavelength range. The visible photon is used for spatially resolved detection because efficient single-photon detector arrays are available. In this way, single-photon imaging in the infrared range can be realized [48].

For example, this method can be used to image objects and structures that are opaque (total absorption) or invisible (total transmission) to visible light. Another advantage of the method is the very low illumination

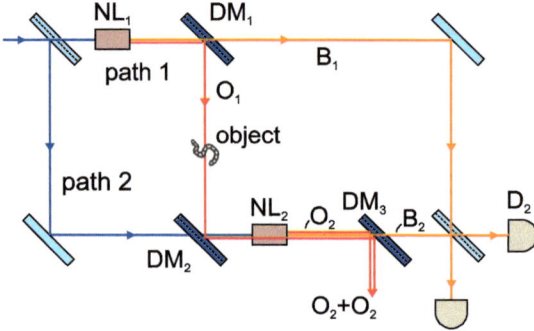

Figure 4.6: Schematic of quantum imaging with undetected photons.

intensity to which the object is exposed. This can be advantageous for microscopic imaging of living cells or microorganisms.

Quantum ghost imaging, as described above, is based solely on spatial and temporal correlations between two photons. True quantum effects do not play a central role. Therefore, comparable methods inspired by quantum ghost imaging have also been realized with classical light [49].

Quantum imaging with undetected photons

In quantum ghost imaging, the photons in the object path are still detected (with the bucket detector). Newer approaches allow imaging without measuring the photons interacting with the object to be imaged [50]. The goal here is to extend the wavelengths that can be used to areas where single photon detectors are unavailable.

The starting point for understanding the setup shown in Fig. 4.6 is the quantum bomb test from Section 4.2. It is basically a Mach-Zehnder interferometer set up in a way that, due to destructive interference, photons are only detected by one of the two detectors. Interference can only occur if the two paths in the interferometer are indistinguishable (basic rule 4). If the paths are distinguishable, photons are detected by both detectors. This is used to image the object: The absence of destructive interference indicates the presence of the object.

Instead of the "classical" Mach-Zehnder interferometer, a nonlinear interferometer setup is used in which pairs of photons are generated by parametric fluorescence (spontaneous parametric down-conversion) at two different locations (in the nonlinear crystals NL_1 and NL_2). In each case, one photon of the first pair is superposed with the corresponding photon of the other pair so that interference can occur. Specifically, the experiment consists of the following steps:

1. The pump laser beam is split into two paths, path 1 and path 2.
2. In path 1, spatially entangled photon pairs of different wavelengths are generated in a nonlinear crystal NL_1 (shown in orange and red in Fig. 4.6).

3. The photon pairs are separated into photons for imaging (B_1) and photons interacting with the object (O_1). This is done with a dichroic mirror (DM_1) that selectively reflects only one color (red) and is transparent for the other color (orange).
4. In path 2, entangled photon pairs are also generated in an identical nonlinear crystal and separated in the same way (B_2 and O_2) using the dichroic mirror (DM_2).
5. The imaging photons from the two paths (B_1 and B_2) are superposed at the second beamsplitter so that they become indistinguishable.
6. Similarly, the two object photons (O_1 and O_2) are also superposed at the third dichroic mirror (DM_3) so that they are also indistinguishable, unless an object is introduced in the path of O_1.
7. In this scenario (with no object), the imaging photons interfere destructively at detector D_2 (so that the signal of D_2 is zero).
8. The other two photons (O_1 and O_2) also interfere, but are not detected.
9. If, on the other hand, an object is in the path of O_1, the two paths O_1 and O_2 become distinguishable. No interference occurs – not even for the imaging photons B_1 and B_2. No destructive interference occurs at detector D_2; photons are detected.

With this method, the presence of an object can be inferred from the absence of interference of the imaging photons. In this case, the imaging photons never interacted with the object, and the photons that did interact with the object are not measured. The advantage is that different wavelengths can be used for the photons of the correlated pairs. For the imaging photons, a convenient wavelength can be used. Since the object photons are not detected, it is not necessary to have single-photon detectors for them. Compared to the previously described methods, new wavelength ranges can be explored.

4.4 Interferometry at the quantum limit

Standard Quantum Limit

In interferometric measurements, the goal is to detect a phase difference between the two arms. This can be done by adjusting the interferometer – as in the discussion of the quantum bomb test – in the initial state so that the intensity at one of the outputs is zero. If now an object is introduced that changes the phase in one of the arms due to its refractive index, the intensity at that output will become nonzero. The magnitude of the phase shift is indicated by the intensity of the outgoing light.

However, due to the probabilistic nature of quantum physics, the intensity at the output of the interferometer fluctuates. The photons emitted there do not arrive at the detector as a uniform stream but fluctuate. These fluctuations in the number of photons are called *shot noise*. Shot noise scales with $\frac{1}{\sqrt{N}}$, where N is the average number of incoming photons. This noise limit, which limits the accuracy of the measurement, is called the *Standard Quantum Limit* (SQL).

Shot noise occurs whenever energy is transmitted in discrete "packets" whose arrival is determined by a random process – historically first studied for electrons in a vacuum tube. Such random processes are mathematically described by Poisson distributions whose standard deviation scales with $\frac{1}{\sqrt{N}}$ where N is the mean. To reduce the relative error, the measurement can be repeated N times, or the intensity can be increased to N times. This will reduce the error proportionally to $\frac{1}{\sqrt{N}}$. This is the standard quantum limit, which, in the case of photons, is due to the discrete nature of light quanta.

NOON states

However, quantum mechanics allows a higher accuracy. Instead of considering N independent single events, one increases the intensity by a single event with N times the intensity. This is the idea behind *NOON-states* (written with zeros and pronounced like *noon*). They are generally represented by the expression [51]

$$\frac{1}{\sqrt{2}}(|N,0\rangle + |0,N\rangle). \tag{4.2}$$

To generate NOON states, one must achieve that after the beamsplitter, there are exactly 0 photons in one arm of the interferometer and exactly N in the other arm – or exactly the opposite. Such a state has two advantages for interferometry:
1.   It collects N times the phase difference on its way through the interferometer. This increases the sensitivity of the interferometer.
2.   In a detection event, it is not N individual photons that are detected at the detector, each with a certain probability, but either N photons as a whole or none at all. As of this all-or-nothing nature of the NOON states, the scale of fluctuations with $\frac{1}{N}$. This translates into an improvement in measurement accuracy of $\frac{1}{\sqrt{N}}$ over the standard quantum limit.

The scaling of the measurement accuracy with $\frac{1}{N}$ is called *Heisenberg limit*; it is the limit of what is possible in quantum mechanics [52].

 Hong-Ou-Mandel effect
To generate an NOON state for $N = 2$, the *Hong-Ou-Mandel (HOM) effect* can be used. This effect occurs when two photons are sent to a beamsplitter from two directions. At the beamsplitter, they are superposed so that the transmitted and reflected states of both photons are indistinguishable. If this is the case, they interfere so that they are in a common entangled state behind the beamsplitter:

$$\frac{i}{\sqrt{2}}(|2,0\rangle + |0,2\rangle). \tag{4.3}$$

When measured, both photons are always detected in the same arm of the interferometer; the case where one photon is detected in one arm and the other photon in the other arm does not occur.

NOON states are entangled states. They are fragile and vulnerable to decoherence. Their creation and handling in experiments become difficult for $N > 2$. NOON states up to $N = 5$ have been demonstrated in experiments so far.

4.5 Quantum logic spectroscopy

Precision spectroscopy of individually trapped and cooled ions and atoms has long been an established and highly successful area of metrology. High spectroscopic resolution is achieved by *laser cooling*. The trapped atom or ion undergoes cycles of targeted excitation and de-excitation of internal energy levels. The laser frequency is chosen so that during each cycle, the ion loses kinetic energy – the ion is slowed down, and its thermal motion in the trap potential becomes increasingly slower. In this way, trapped ions can be cooled down to the microkelvin range. This is a significant advantage for spectroscopy because it reduces the Doppler frequency shifts caused by the ion's motion, which otherwise limits the accuracy of spectroscopic measurements. There are a number of requirements for the ions used in order for the method to work:
(a) a suitable spectroscopic transition, i. e., a narrow transition with a long excited state lifetime,
(b) a fast transition for the cooling sequence in laser cooling,
(c) the initial state must be easy to prepare,
(d) efficient state detection.

By far, not all atomic and ionic species meet these requirements. In order to extend the range of application to other species, *quantum logic spectroscopy* has been developed. Here, methods originally used to realize quantum gates are made usable for spectroscopic purposes [53].

The ion to be studied only has to fulfill the first of the four requirements. In addition to this ion (the *spectroscopy ion*), a second ion (the *logic ion*) is used, a well-controllable ion that meets the other three requirements. The basic idea of quantum logic spectroscopy is to transfer information about the internal state of the spectroscopy ion to the logic ion and to read it from there [54, 55]. As a further advantage, the spectroscopy ion is also cooled via the logic ion.

The spatial vibrations of the two ions are common degrees of freedom. They are used as a means to transfer the state from one ion to the other. The two ions experience the common trap potential. Due to their charge, they repel each other. As a result, oscillations around the equilibrium state (where the ions are separated by a few micrometers) can be excited. According to quantum mechanics, these oscillations have discrete energy values. To bring the system from the ground state to the first excited state, a precisely defined amount of energy is required. This is exploited in quantum logic spectroscopy.

Procedure of quantum logic spectroscopy
The diagrams in Fig. 4.7 show the states of the ions involved in the process. The states of the spectroscopy ion are shown on the left and those of the logic ion on the right. In addition to the internal energy levels, there are the vibrational degrees of freedom already mentioned. The vibrational ground state and the first excited state, which are relevant for the scheme, are labeled $n = 0$ and $n = 1$ in Fig. 4.7. The vibrational states

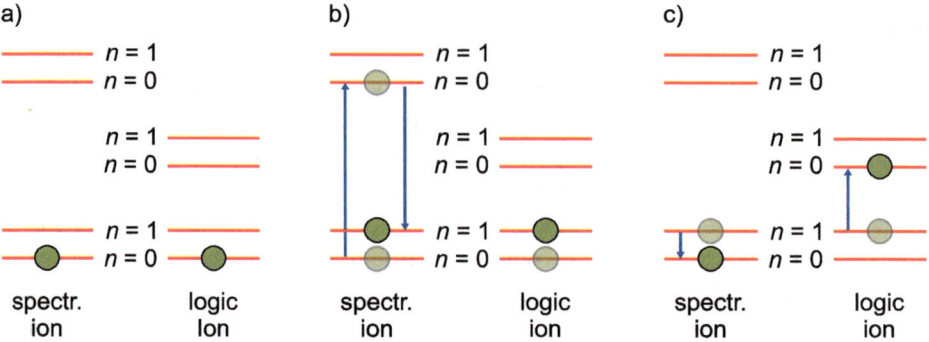

Figure 4.7: Quantum logic spectroscopy.

are common to both ions. Therefore, both ions must necessarily be in the same state, i. e. both in $n = 0$ or both in $n = 1$. This is physically obvious but difficult to represent in diagrams where the energy levels are drawn separately for each ion. The procedure of quantum logic spectroscopy is as follows:

(a) Laser cooling brings both ions to the ground state, both for the internal energy levels and the vibrational degrees of freedom (Fig. 4.7(a)).

(b) The transition to being examined in the spectroscopy ion is excited by a laser pulse without exciting vibrations (Fig. 4.7(b)). The ions remain in the vibrational ground state $n = 0$. The second pulse returns the spectroscopy ion to the internal ground state. The frequency of this pulse is slightly lower than that of the first one, so the spectroscopy ion cannot return to the initial state. However, the transition is possible if the $n = 1$ state of the vibrational degrees of freedom is simultaneously excited. This automatically affects both ions, i. e., the logic ion is now also in the $n = 1$ state.

(c) With a third pulse, the logic ion is brought into the excited internal state, and the common oscillation is brought back to the ground state $n = 0$ (Fig. 4.7(c)). Now the readout can take place.

The sequence proceeds in this way if the excitation of the spectroscopy ion is successful. If this was not the case, e. g., because the first laser pulse did not hit the resonance frequency of the spectroscopy ion, the system remains in the initial state. In the first case, the logic ion is in the excited internal state; in the second case, it is in the ground state. Thus, the internal state of the logic ion indicates whether the excitation of the spectroscopy ion was successful. The logic ion species is chosen in such a way that its state is easy to detect (by excitation to a third internal state, not shown). Thus, the fluorescence intensity of the logic ion provides information about the energy levels of the spectroscopy ion.

Quantum logic spectroscopy is of particular interest for optical atomic clocks because this technique allows the use of ions that have advantageous spectroscopic properties but lack transitions suitable for laser cooling and detection. An example is the Al^+

ion, whose relevant transition is particularly insensitive to external perturbations (electric and magnetic fields, thermal radiation) and has an extremely narrow linewidth of only 8 mHz. An Al^+ quantum logic clock was realized in 2019 at NIST in Boulder [56] with an accuracy of less than 10^{-18}. This corresponds to a deviation of one second in 30 billion years.

5 Quantum information and communication

5.1 Impossible machines

Broad areas of quantum technologies, such as quantum communication and quantum computing, are based on the processing and transmission of information. The special nature of *quantum information* gives it advantages over its classical counterparts. Quantum information is the information stored and transmitted in the states of individual quantum systems. In the following, we will show that quantum information is indeed something new. It is not easily described by the concepts of classical information theory and cannot be realized by information carriers based on classical physics.

The nonclassical nature of quantum information has been most impressively demonstrated by Werner [57, 58], who designed a hierarchy of "impossible machines" that show how classically plausible assumptions about quantum information lead to unacceptable consequences. The machines are such that if we have a "stronger" one, we can build the "weaker" one. It turns out that we can violate relativistic causality even with the weakest machine of all, the correlation telephone. If we do not want to accept this, then all stronger machines are excluded.

The classical translator

All the machines described below do not exist in reality. However, it is easy to imagine them and consider what would follow from their existence. The first machine, the classcial translator, would be – if it existed – the strongest of all, although its description sounds the most harmless. It is shown symbolically in Fig. 5.1 (top). It receives as input a quantum system and can extract from it, with a single measurement, the complete information about its state (about which nothing else is known beforehand). This process sounds quite innocent because it is taken for granted in classical information theory. In quantum physics, it is not. In general, the state of a quantum system cannot be determined by a single measurement but can only be reconstructed from the results of many measurements. A simple example is the polarization measurement described on p. 72: a single measurement yields a single result, such as a detector click at the + output (cf. basic rule 3). More information about the state of the incoming photons can only be obtained from the statistics of many measurements.

"Complete state information" means that the existence of a classical translator would make it possible to build the classical teleportation machine shown in Fig. 5.1 (bottom). A preparation device P would be able to use this (classical) information to prepare a quantum system that is indistinguishable from the original one for arbitrary measurements. In quantum physics, this requires the knowledge of the state vector or the corresponding density matrix.

https://doi.org/10.1515/9783110717457-005

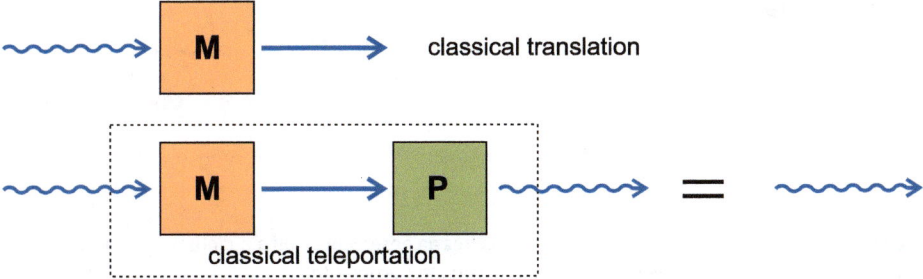

Figure 5.1: Classical translator (top): A single measurement M provides complete information about the state of any quantum system. This allows (via the preparation P) classical teleportation (bottom). Wavy arrows represent quantum systems; straight arrows represent classical information.

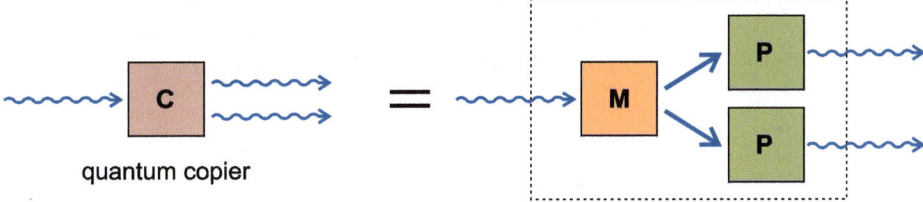

Figure 5.2: A quantum copier (left) takes a single quantum system as input and makes a perfect copy. If classical translation is possible, a quantum copier can be built (right).

Quantum State Tomography: In general, state reconstruction is a difficult problem in quantum mechanics. In *quantum state tomography*, numerous measurements on an ensemble of identically prepared quantum objects are used to reconstruct their state. The measurements have to be done on a large number of different bases. The name of the method comes from its similarity to medical tomography, where the patient is X-rayed from many different directions in order to reconstruct the inside of the body on a computer.

Quantum copier

The quantum copier is a device that takes as input a single quantum system in an arbitrary state and makes a perfect copy of it (Fig. 5.2), indistinguishable from the original by any measurement. The right part of Fig. 5.2 shows that it is possible to construct a quantum copier using a classical translator: Since classical information can be duplicated arbitrarily, it is sufficient to send the information obtained from the classical translator to two preparation devices to make a perfect copy of the original quantum system. Any number of copies can be made in this way.

No-cloning theorem

The statement *"A single quantum cannot be cloned"* summarizes the no-cloning theorem of Wooters and Zurek [59]. It was first explicitly stated in 1982, very late in the history of quantum physics. Today, it is the cornerstone of quantum communication, particularly quantum cryptography, where it guarantees the secu-

perturbation-free measurement

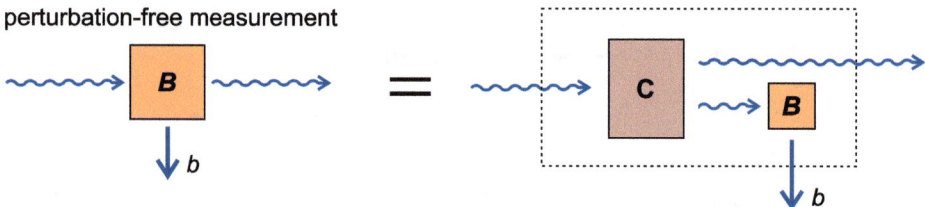

Figure 5.3: A perturbation-free measurement measures an observable B on a quantum system without changing its state (left). It can be performed using a quantum copier (right).

rity of quantum cryptographic protocols against eavesdropping. The no-cloning theorem states that a quantum copier is impossible within the framework of quantum mechanics. It is impossible to perfectly copy a quantum system in an arbitrary and unknown state.

The proof of this theorem is straightforward. For simplicity, we will consider abstract qubit states. A quantum copier would copy the state $|\psi\rangle$ of one qubit to a second qubit that was originally in a state $|a\rangle$. It would thus bring about the following time evolution (the indices label the qubits):

$$|\psi\rangle_1 |a\rangle_2 \longrightarrow |\psi\rangle_1 |\psi\rangle_2. \tag{5.1}$$

In particular, this should be true for the qubit basis states $|0\rangle$ and $|1\rangle$:

$$|0\rangle_1 |a\rangle_2 \longrightarrow |0\rangle_1 |0\rangle_2, \quad |1\rangle_1 |a\rangle_2 \longrightarrow |1\rangle_1 |1\rangle_2. \tag{5.2}$$

Since the superposition principle holds in quantum mechanics, this already determines the time evolution of any superposition state:

$$|\psi\rangle_1 |a\rangle_2 = \left[\alpha |0\rangle_1 + \beta |1\rangle_1\right] |a\rangle_2 \longrightarrow \alpha |0\rangle_1 |0\rangle_2 + \beta |1\rangle_1 |1\rangle_2. \tag{5.3}$$

This is not the desired final state $|\psi\rangle_1 |\psi\rangle_2$. Thus, the superposition principle prevents the perfect copying of any quantum state.

Perturbation-free measurement

An apparatus for perturbation-free measurement measures an observable B on a quantum system in an arbitrary state without perturbing that state (Fig. 5.3). It thus contradicts the Heisenberg indeterminacy relation (in the formulation "no measurement without perturbation" discussed on p. 74). As shown in Fig. 5.3 (right), we can perform such a measurement if we have a quantum copier. We copy the quantum system and perform the measurement on the copy. The original quantum system remains undisturbed.

Joint measurement

Another formulation of the indeterminacy relation discussed on p. 74 states that two complementary observables B_1 and B_2 cannot be measured simultaneously without the measurements interfering with each other: If the measurements are repeated many times, the probability distributions obtained for the joint measurement will differ from

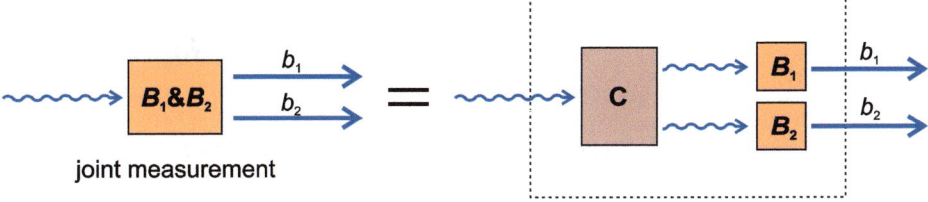

joint measurement

Figure 5.4: A joint measurement of two observables can be performed using a quantum copier.

the distributions measured individually. The polarization components of light in the H/V basis and the +/− basis are an example of complementary observables for which there is an indeterminacy relation.

A hypothetical device for a joint measurement of the observables B_1 and B_2 (denoted $B_1 \& B_2$ in Fig. 5.4) is capable of doing just that. It returns two values b_1 and b_2 for every single measurement so that when repeated many times, all probabilities and statistical distributions are the same as if B_1 and B_2 were measured individually.

As can be seen from Fig. 5.4 (right), a joint measurement device can be constructed with a quantum copier. This is done by copying the original quantum system and performing the measurements of B_1 and B_2 individually on the two identical quantum systems with conventional measuring devices.

Correlation telephone

The correlation telephone (also called Bell's telephone) is a device that allows communication by measurement. In quantum mechanics, this is not possible in principle, but as we will see, a corresponding scheme could be realized if one had an apparatus for joint measurement. If such a device existed, it would be possible to transmit messages instantaneously, i. e., faster than the speed of light. Such a possibility would be in sharp contrast to special relativity and would raise causality problems.

If a joint measurement device were possible, it could be used to transmit information by measurement in the following way (Fig. 5.5): Alice and Bob share a sequence of entangled photon pairs. Alice wants to send a message consisting of zeros and ones to Bob. She would do this as follows: If she wants to send a 0, she chooses the first measurement device and measures the observable A_α. If she wants to send a 1, she chooses the second measurement device and measures A_β. Bob has a device for joint measurement of the observables B_γ and B_δ. He decodes the message as follows:

- If the two measurements b_γ and b_δ are equal (i. e. either both equal +1 or both equal −1), then he interprets that as 0.
- If the two measurements are different, then he interprets that as 1.

The following box shows that messages could indeed be sent using this method, although only by exploiting statistics.

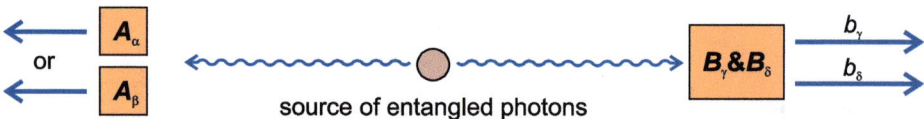

Figure 5.5: A correlation telephone could be constructed using a joint measurement apparatus and entangled photons.

ⓘ **How the correlation telephone works**

A device for the joint measurement of two complementary variables does not exist according to quantum mechanics, so it cannot be described by the usual formalism of quantum physics. Therefore, the results have to be derived indirectly; they can be traced back with a clever calculation to the correlation functions we have already encountered in Section 3.14 in connection with the Bell-CHSH inequality [57].

We first consider the case where Alice wants to send a 0 and measures the observable A_α. The probability that Bob's two measurements b_γ and b_δ match, i. e. that he interprets the signal correctly, is the sum of the probability that they are both equal to +1 and the probability that they are both equal to −1, again summed over Alice's possible measurements a_α:

$$\sum_{a_\alpha}\left[p(a_\alpha, b_\gamma = +1, b_\delta = +1) + p(a_\alpha, b_\gamma = -1, b_\delta = -1)\right]. \tag{5.4}$$

This expression can be formally written as:

$$\sum_{a_\alpha, b_\gamma, b_\delta}\left|\frac{b_\gamma + b_\delta}{2}\right| |a_\alpha| \, p(a_\alpha, b_\gamma, b_\delta), \tag{5.5}$$

where the sum now extends over all variables. The first term ensures the condition $b_\gamma = b_\delta$, while $|a_\alpha|$ is always 1 and will be useful later.

If Alice wants to send a 1, she measures the observable A_β. Bob interprets the signal correctly if his two measurements b_γ and b_δ are different. The probability for this can be written analogous to the first case:

$$\sum_{a_\beta, b_\gamma, b_\delta}\left|\frac{b_\gamma - b_\delta}{2}\right| |a_\beta| \, p(a_\beta, b_\gamma, b_\delta). \tag{5.6}$$

We now assume that zeros and ones occur equally often in Alice's signal. The total probability p_{ok} that the signal sent by Alice will reach Bob correctly is the weighted sum of the two terms:

$$p_{ok} = \frac{1}{2}\sum_{a_\alpha, b_\gamma, b_\delta} |a_\alpha|\left|\frac{b_\gamma + b_\delta}{2}\right| p(a_\alpha, b_\gamma, b_\delta) + \frac{1}{2}\sum_{a_\beta, b_\gamma, b_\delta} |a_\beta|\left|\frac{b_\gamma - b_\delta}{2}\right| p(a_\beta, b_\gamma, b_\delta).$$

Since $|a| \geq a$ is always true, we can write:

$$p_{ok} \geq \frac{1}{4}\sum_{a_\alpha, b_\gamma, b_\delta} a_\alpha \, (b_\gamma + b_\delta) \, p(a_\alpha, b_\gamma, b_\delta) + \frac{1}{4}\sum_{a_\beta, b_\gamma, b_\delta} a_\beta \times (b_\gamma - b_\delta) \, p_2(a_\beta, b_\gamma, b_\delta).$$

$$= \frac{1}{4}\sum_{a_\alpha, b_\gamma} a_\alpha b_\gamma \, p(a_\alpha, b_\gamma) + \frac{1}{4}\sum_{a_\alpha, b_\delta} a_\alpha b_\delta \, p(a_\alpha, b_\delta)$$

$$+ \frac{1}{4}\sum_{a_\beta, b_\gamma} a_\beta b_\gamma \, p(a_\beta, b_\gamma) - \frac{1}{4}\sum_{a_\beta, b_\delta} a_\beta b_\delta \, p(a_\beta, b_\delta).$$

The terms in the last equation are just the classical definition of the correlation function between the two variables that appear (in each of the terms, the third variable has been summed over):

$$p_{ok} \geq \frac{1}{4} \left[\langle A_\alpha B_\gamma \rangle + \langle A_\alpha B_\delta \rangle + \langle A_\beta B_\gamma \rangle - \langle A_\beta B_\delta \rangle \right]. \tag{5.7}$$

This is the condition for Alice to be able to send signals to Bob by choosing a measurement device, provided that Bob is able to make a joint measurement of B_γ and B_δ. Note that the expression in parentheses simply corresponds to the combination of correlation functions that occurs in the Bell-CHSH inequality (3.136).

The probability p_{ok} is greater than the probability of guessing (which is $\frac{1}{2}$) exactly if the expression in parentheses is greater than 2 – that is, if the Bell-CHSH inequality is violated. This is a remarkable result. Two conditions are necessary for a correlation telephone to work: the possibility of joint measurement of complementary observables and the violation of the Bell-CHSH inequality. Thus, the correlation telephone would not work either in a purely classical world or in the framework of quantum mechanics. In the purely classical world, the Bell-CHSH inequality is not violated; in quantum mechanics, joint measurement is not possible. Thus, faster-than-light communication is impossible in both cases, but for different reasons.

Quantum information and the hierarchy of impossible machines

The sequence of impossible machines presented in this section is designed in such a way that each subsequent machine can be constructed from the previous ones. Thus, they form a hierarchy – from the classical translator, via the quantum copier, the perturbation-free and joint measurement, to the correlation telephone. The classical translator is the strongest of the devices because all others can be constructed from it. The weakest of all machines, which can be constructed from all others, is the correlation telephone – but it would have drastic effects if it could be realized.

If we insist on relativistic causality (which excludes sending signals faster than the speed of light), then the correlation telephone cannot exist, and so neither can any of the stronger machines. This is not an additional restriction we have to impose on quantum mechanics. As we have seen, all the devices are excluded anyway in quantum mechanics. The hierarchy of impossible machines has a different purpose: it illustrates the novel character of quantum information and shows us that classical information and quantum information are very different concepts.

5.2 Quantum cryptography

In the digital age, encrypted data transmission is more important than ever. Today, every web page is transmitted in encrypted form, recognizable by the "https" in the web address (s = secure). Clicking on the lock icon in the address bar provides more detailed information about encryption and data security during transmission. Encrypted data transmission is important today not only in online commerce but in all areas of business and society. This is especially true for critical infrastructure, many industrial applications, the financial sector, medical data, and the military sector. Security of data in cloud storage is also an important application of encryption methods.

In *cryptography*, the role of quantum technologies is twofold: on the one hand, quantum computers have the potential to break some of the most widely used encryption schemes currently in use. On the other hand, with quantum cryptography, they provide novel quantum-based methods for secure communication. We first consider the threat of quantum algorithms to currently used cryptographic techniques before discussing quantum cryptography in the following sections.

The threat of quantum computers to encrypted communication

Today, the *RSA algorithm* is most commonly used for encrypted communication. It is based on the fact that there is no known classical algorithm capable of efficiently factorizing large numbers. The computation time required by classical algorithms scales exponentially with the number of digits of the numbers to be factorized. The security of the RSA algorithm is based on the fact that it is practically impossible to factor sufficiently large numbers in an acceptable amount of time (more on this in Section 6.6).

In 1994, the *Shor algorithm* was introduced, which would allow the factorization of large numbers in polynomial time if it could be implemented on a quantum computer (cf. Section 6.6). This would eliminate the basis for the RSA method. Parties possessing a powerful quantum computer would be able to decrypt and read communications encrypted using the RSA algorithm. With the prospect of functional quantum computers in the near- or mid-term future, there is growing concern about the security of conventionally encrypted data. This also applies to sensitive data sent today or stored in the cloud, which can be intercepted today but decrypted in the future ("harvest today, decrypt tomorrow").

If only one of the classical encryption algorithms was affected by the threat of quantum computers, it would be easy to switch to other algorithms. For instance, the widely used *AES algorithm* is not based on the factorization of large numbers. However, its decryption can be traced back to the solution of a system of polynomial equations. Here, another quantum algorithm, the *HHL algorithm* (cf. Section 6.8), has the potential for an efficient solution – although many details are still unsettled. The *Grover algorithm* for searching unordered databases (cf. Section 6.3) can also speed up algorithms like the number field sieve (but, unlike Shor's algorithm, not exponentially). Overall, quite a number of classical encryption algorithms are affected by the threat of quantum computers, especially the *public-key cryptography algorithms* (RSA, Diffie-Hellman, ElGamal, ECIES, DSA, ECC). However, the full extent of the challenge posed by quantum algorithms cannot yet be conclusively foreseen [21].

The further development of classical cryptography is a possible way out of this situation. In *post-quantum cryptography*, new classical encryption algorithms are explored that are secure against decryption attacks from quantum computers. As part of the Post-Quantum Cryptography Standardization Process of the US National Institute of Standards and Technology (NIST), such algorithms have been tested for their vulnerability

Figure 5.6: Cipher disk with alphabets arranged in opposite directions in the setting A → N.

in a multi-stage process since 2017. Only one algorithm for public-key encryption and three for digital signatures have made it to the fourth round (2022).

Cryptographic keys and one-time pads

The second way to ensure secure data transmission is *quantum cryptography*. Here, the fragile nature of quantum information is used to protect transmitted messages from eavesdroppers. The most common methods of quantum cryptography do not transmit the messages themselves but generate cryptographic keys that are used to securely encrypt the message to be transmitted. The encrypted message can then be transmitted over a classical and potentially insecure channel.

For the background of this concept, we have to go back to one of the oldest encryption methods used by Julius Caesar for secret communication. In this method, all letters are shifted by a number of positions in the alphabet, such as A → D, B → E, and C → F. The easiest way to do this is with a cipher disk, as shown in Fig. 5.6. However, the *Caesar cipher* is anything but secure: In every language, some letters occur more frequently than others (in English, these letters are E, A, R, I, O, T). Therefore, in longer texts, a frequency analysis can be performed, which provides a starting point for decryption. All encryption algorithms that assign the same character in the encrypted text to the same character in the plaintext can be broken by frequency analysis.

A provably secure encryption scheme is based on the use of a *one-time pad* (one-time key). This is a random string that is at least as long as the text to be encrypted and is used only once. Fig. 5.7 explains the procedure. The one-time pad is the random string "HPZVME". Both Alice and Bob must know this key.

Each letter of the message ("SECRET" in Fig. 5.7) is encrypted by Alice using a different position of the cipher pad specified by the entries in the one-time pad. The first letter is encrypted with the position A → H, the second with A → P, and so on. The encrypted message "PLXFIL" generated in this way can be distributed over a public

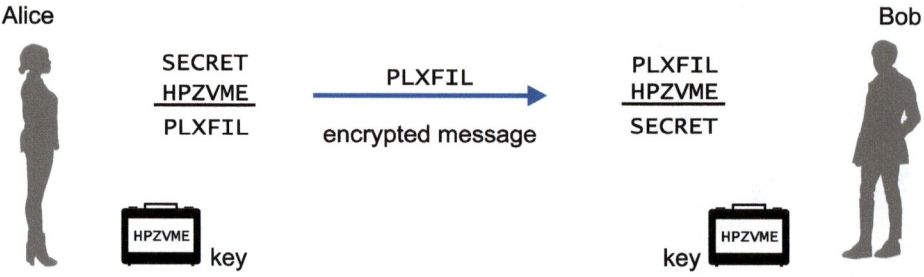

Figure 5.7: Encrypting a message with a one-time pad.

medium (e. g., telephone, radio, or internet) because it is of no value to an eavesdropper without the key. Only the recipient, Bob, can use the key and a cipher pad to decrypt the message by performing the reverse operations as Alice. In reality, of course, it is not the letters of the alphabet that are encrypted but their binary representations. As the same key is used for both encryption and decryption, the scheme is called a *symmetric method*.

One-time-pad cryptography is provably secure from an information-theoretic point of view, provided that the key is as least at long as the plaintext message, randomly generated, kept secret, and never reused. The problem is the practical side of key distribution. Alice and Bob can physically exchange the key (e. g., in the form of hard disks filled with random codes). In reality, this is not very convenient; another solution must be found. Thus, using one-time pads reduces the problem of eavesdropping-proof communication to a different but simpler problem: secure key distribution. This is the task of quantum cryptography.

Key distribution with polarized photons

The oldest and still paradigmatic method of quantum key distribution was developed by Bennett and Brassard in 1984 and is therefore called *BB84 protocol* [60]. It uses the polarization degrees of freedom of single photons to encode information. For example, in the H/V basis (cf. p. 66), the value 0 could be encoded as "horizontally polarized" (H) while the value 1 could be encoded as "vertically polarized" (V). If Alice is able to generate single photons with a defined polarization, and Bob has a single photon detector and a polarizing beam splitter with which he can measure the polarization in the H/V basis, then Alice and Bob can transmit information in this way.

However, this method is not secure. An eavesdropper (Eve) could intercept a transmitted photon, measure its polarization, and send a new photon with the same polarization to Bob (intercept-resend attack; Fig. 5.8). If Eve measures and transmits in the same basis as Alice and Bob, nothing in Bob's measurement results will indicate her presence (assuming perfect detectors and photon sources).

Therefore, Alice and Bob have to use multiple bases. In addition to the H/V base, they can use the +/− base, in which the axes of the polarizing beam splitter are rotated

Alice   Bob

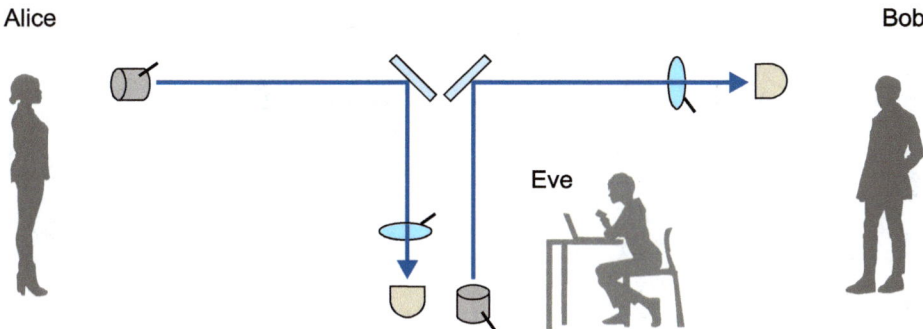

Eve

Figure 5.8: The eavesdropper, Eve, intercepts photons and resends them. If only one basis is used, the presence of Eve cannot be detected by Alice and Bob.

by 45°. There is a indeterminacy relation between the two quantities; they cannot be prepared simultaneously without scattering (cf. p. 71). A polarization state with a fixed value in the H/V basis gives completely random results when measured in the +/− basis. Alice and Bob use the indeterminacy relation to make their message inaccessible to Eve.

> Quantum cryptography is the secure distribution of cryptographic keys between two parties (Alice and Bob).

The BB84 protocol

In the BB84 protocol, both Alice and Bob randomly and independently choose their bases for preparation or measurement for each run. This is done individually for each photon – without prior agreement. Afterwards, they have to exchange information about the bases they use. The BB84 protocol consists of the following steps (Fig. 5.9):

1. Alice has a sufficiently long sequence of binary random numbers that she wants to send to Bob – from which the cryptographic key will later be generated.
2. She randomly chooses between H/V or +/− basis for transmitting a bit. She makes a note of the basis and sends an appropriately polarized photon to Bob (encoding the values 0 and 1 as described above).
3. Bob randomly decides in which basis to analyze the incoming photon (H/V or +/−). He notes down the measurement result and the basis used.
4. After more than twice, as many photons have been sent as the final length of the key should be, Alice and Bob contact each other and exchange their notes about the chosen basis. This can be done over an insecure connection (e. g., telephone or internet). Eavesdropping on this information is harmless.
5. Alice and Bob cross out all the bits where they used different bases (in Fig. 5.9 the third and fifth bit). This affects about half of the entries in their list. The rest of the bits, where they used the same basis, were correctly reconstructed by Bob. As

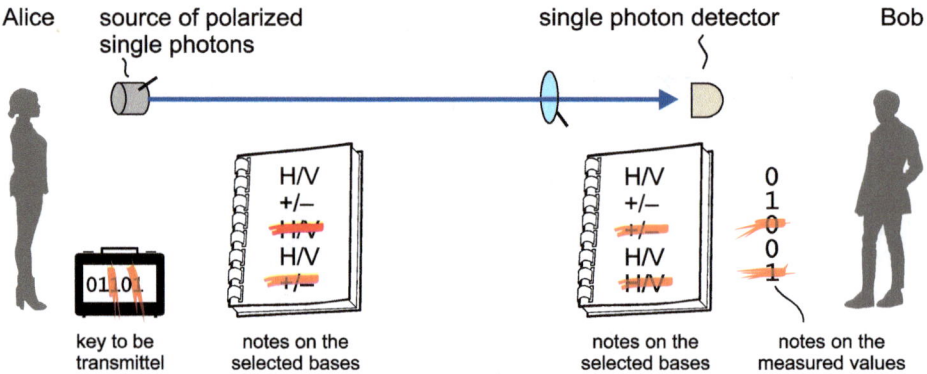

Figure 5.9: Principle of the BB84 protocol. Bob and Alice choose between H/V basis and +/− basis for each photon. In addition, a classical communication has to be done.

a result, these bits (in Fig. 5.9 "010") are now available to both Alice and Bob. They can now be used as the key for the one-time-pad encryption.

The protocol entails that about 50 % of the bits transmitted have to be discarded after being detected. This does not matter too much, though, because it is not a message that is being transmitted, just a cryptographic key made up of random bits. A random sequence in which half of the bits are deleted at random is still a random sequence, so it can be used as a one-time pad.

Detection of eavesdroppers

Using the BB84 protocol, Eve's attack (Fig. 5.8) can now be revealed. This is because she has no other option but to guess the measurement basis for each photon. In half of the cases, she coincides with Alice's choice of the basis. She can then extract the transmitted information as before and send an identically prepared photon to Bob. In the other half of the cases, however, she is wrong in her choice of measurement basis. Then she sends a photon prepared on the wrong basis to Bob; he receives a random result. The probability that it matches Alice's original bit is $\frac{1}{2}$.

Thus, on average, 25 % of the bits sent by Alice arrive at Bob with the wrong value. This reveals the presence of Eve. Alice and Bob must sacrifice part of their key and publicly compare a certain number of bits sent. If the error rate is 25 % or higher, they know they are being intercepted and can act accordingly.

The basis for Eve's detection in this scenario is the indeterminacy relation in the "no measurement without perturbation" version. Any attempts by Eve to extract information from Alice's signal will perturb the signal, thus revealing Eve's attack. In another scenario, conceivable in classical physics, Eve "steals" the photon sent by Alice by copying it. However, as the no-cloning theorem (p. 117) shows, copying an arbitrary unknown quantum state is impossible. Thus, Eve also remains unsuccessful in this case.

Other protocols and practical implementation of quantum cryptography

The simple thought experiments with Alice, Bob, and Eve are mainly used to illustrate the basic ideas of quantum cryptography. They show how the principles of quantum mechanics can be used to enable the secure transmission of information.

Generally, a distinction is made between protocols based on the indeterminacy relation (such as BB84) and those based on entanglement. The prototype of entanglement-based protocols, the E91 protocol, was proposed by Ekert in 1991 [61]. Instead of single photons, it uses entangled photon pairs distributed between Alice and Bob. Very similar to the BB84 protocol, they perform their measurements in various bases and then publicly exchange information about the bases used. They establish the integrity of their data by violating Bell's inequality. Interference by an eavesdropper would result in the inability to detect a violation of Bell's inequality.

For practical implementation of the basic ideas, the protocols need to be refined and adapted to the practical conditions [62]. For example, the desired high data rates are hardly compatible with controlled single-photon states with exactly one photon in the pulse. It is much easier to work with attenuated laser pulses, where the pulse contains a superposition of 0, 1, 2 or more photons. The drawback that one of these photons could, in principle, be redirected for eavesdropping is accepted in favor of better practicability.

Removing photons unnoticed is part of a scenario called the *photon-number-splitting attack*, in which Eve blocks all one-photon states and allows only the two-photon states to pass to Bob. Of every two-photon state, he keeps one photon and – after waiting for the public communication between Alice and Bob – decrypts the message. To prevent this, the idea of *decoy states* was introduced [63]. The BB84 protocol is extended to include additional "decoy pulses" in which Alice intentionally sends pulses with increased photon numbers. The presence of Eve is revealed by lower loss rates for these pulses. Alice and Bob can detect this by public comparison.

In practice, cryptographic protocols must also be secured against *side-channel attacks*. These are attacks that do not target the algorithm, but rather the physical device used to implement it and extract information from it. Historically, the first side-channel attack on a quantum cryptography protocol could have happened right at the first experimental implementation of the BB84 protocol [64] in 1991. In that experiment, Pockels cells were used to set the polarization basis for Alice and Bob. Their driver electronics produced such a loud click that a nearby eavesdropper could have obtained information about the bases just by listening.

In real-world environments, side-channel attacks on the hardware are a real risk. Varying processing times, power consumption, optical properties of the components, and even implementation flaws could, in principle, reveal the information to be transmitted. All of these potential vulnerabilities must be analyzed and controlled before quantum cryptographic systems can be considered secure [21].

On the way to practical implementation of quantum cryptography, many technical challenges arise. Of fundamental importance is the type of transmission channel (optical fiber or free space). Fiber optics is used at telecom frequencies (1300 nm or 1550 nm) where optical fibers have small transmission losses. At these frequencies, however,

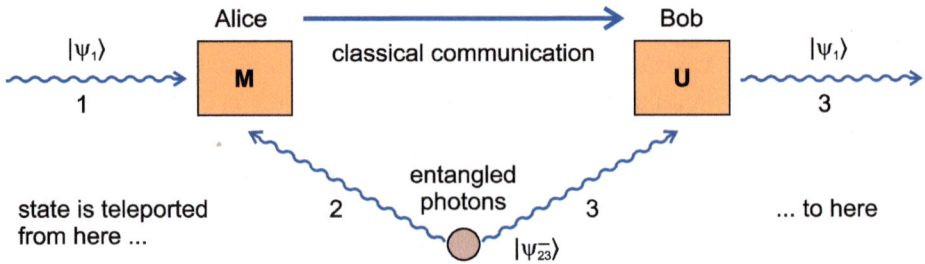

Figure 5.10: Principle of quantum teleportation.

single-photon detectors are inherently more noisy and less efficient than at shorter wavelengths. For free space communications, e. g., with satellites, frequencies in the near-infrared (800–850 nm) are used because there is little absorption from the atmosphere, and there are low-noise single-photon detectors available. For communication over longer distances, one has to deal with the problem of absorption in the transmission channels. In optical fibers, photons can travel, on average, only a few tens of kilometers before they are absorbed. To circumvent this problem, *quantum repeaters* are needed, which allow the transmission of quantum states over longer distances (cf. Section 5.4).

5.3 Quantum teleportation

Quantum teleportation is a central fundamental operation in quantum communication. It allows the transfer of an unknown quantum state from one qubit to another. It requires an entangled pair of photons and a classical communication channel. The concept of quantum teleportation was introduced by Bennett et al. in 1993 [7] and was realized experimentally in 1997 [9]. Contrary to what the name suggests, quantum teleportation does not involve the transport of a physical object (there is no "beaming" in the sense of science fiction). Only the quantum mechanical state information is transferred from one place to another, from Alice to Bob.

The basic situation is shown in Fig. 5.10. It is an extension of the classical teleportation of Fig. 5.1. Alice has a qubit in the state $|\psi_1\rangle$, which is unknown to her, and she wants to send this state to Bob. As we saw in Section 5.1, she cannot determine the qubit state with a single measurement. However, compared to Fig. 5.1, Alice and Bob have an additional common resource: an entangled photon pair (qubit 2 and 3), which is in the singlet state $|\Psi^-\rangle$ (cf. Eq. (3.127)). This additional resource makes it possible to transfer the state $|\psi_1\rangle$ from Alice to Bob. The process requires a measurement, a message sent from Alice to Bob via a classical channel (e. g. by phone), and a qubit operation by Bob. The result of these manipulations is that the state $|\psi_1\rangle$ is destroyed at Alice's place and appears at Bob's. It is transferred to Bob's part of the entangled pair (qubit 3). This transfer of state from qubit 1 to qubit 3 is called quantum teleportation.

ℹ️ In the following consideration, we need the Bell basis states from Eq. (3.127). We write them down again in the notation used here:

$$|\Phi^\pm\rangle = \frac{1}{\sqrt{2}}(|00\rangle \pm |11\rangle), \quad |\Psi^\pm\rangle = \frac{1}{\sqrt{2}}(|01\rangle \pm |10\rangle). \tag{5.8}$$

The reverse transformation is:

$$\left.\begin{array}{c}|00\rangle \\ |11\rangle\end{array}\right\} = \frac{1}{\sqrt{2}}(|\Phi^+\rangle \pm |\Phi^-\rangle), \quad \left.\begin{array}{c}|01\rangle \\ |10\rangle\end{array}\right\} = \frac{1}{\sqrt{2}}(|\Psi^+\rangle \pm |\Psi^-\rangle). \tag{5.9}$$

Quantum teleportation procedure

In detail, the teleportation is done in the following steps:

1. *Initial states:* At the beginning, Alice has qubit 1, whose state she does not know, and which we generally write as

$$|\psi_1\rangle = \alpha |0_1\rangle + \beta |1_1\rangle. \tag{5.10}$$

A pair of entangled photons (qubit 2 and qubit 3) in the singlet state

$$|\Psi_{23}^-\rangle = \frac{1}{\sqrt{2}}(|0_2 1_3\rangle - |1_2 0_3\rangle) \tag{5.11}$$

is shared between Alice and Bob. The total state of all three qubits is:

$$\begin{aligned}|\psi_{123}\rangle &= |\psi_1\rangle \, |\Psi_{23}^-\rangle \\ &= \frac{1}{\sqrt{2}}[\alpha |0_1\rangle + \beta |1_1\rangle] \times \left[\frac{1}{\sqrt{2}}(|0_2 1_3\rangle - |1_2 0_3\rangle)\right] \\ &= \frac{1}{\sqrt{2}}[\alpha |0_1 0_2 1_3\rangle - \alpha |0_1 1_2 0_3\rangle + \beta |1_1 0_2 1_3\rangle - \beta |1_1 1_2 0_3\rangle]. \end{aligned} \tag{5.12}$$

2. *Bell measurement by Alice:* Alice now performs a *Bell measurement* on her two qubits 1 and 2. It is important to emphasize that *not* each qubit is measured individually, but the measurement is performed on both qubits simultaneously with respect to the Bell basis (5.8). To understand the result of this measurement, we first use Eq. (5.9) to transform the states of qubits 1 and 2 in Eq. (5.12) into the Bell basis:

$$\begin{aligned}|\psi_{123}\rangle = \frac{1}{2}[&|\Phi_{12}^+\rangle (-\beta |0_3\rangle + \alpha |1_3\rangle) + |\Phi_{12}^-\rangle (\beta |0_3\rangle + \alpha |1_3\rangle) \\ &+ |\Psi_{12}^+\rangle (-\alpha |0_3\rangle + \beta |1_3\rangle) + |\Psi_{12}^-\rangle (-\alpha |0_3\rangle - \beta |1_3\rangle)]. \end{aligned} \tag{5.13}$$

So far, this equation is only the result of a formal transformation. Physically, the following happens: In the measurement that Alice performs on qubits 1 and 2, one of the four possible outcomes is realized (each with probability $\frac{1}{4}$). After the measurement and the associated state reduction, Alice's system (qubits 1 and 2) is in one

Table 5.1: The four possible results of the Bell measurement.

Measurement result at qubit 1 and 2 (Alice)	$\Rightarrow$ State of qubit 3 (Bob)
$\|\Psi_{12}^-\rangle$	$-\begin{pmatrix} a \\ \beta \end{pmatrix} \equiv -\|\psi_1\rangle$
$\|\Psi_{12}^+\rangle$	$\begin{pmatrix} -a \\ \beta \end{pmatrix} = \begin{pmatrix} -1 & 0 \\ 0 & 1 \end{pmatrix}\|\psi_1\rangle$
$\|\Phi_{12}^-\rangle$	$\begin{pmatrix} \beta \\ a \end{pmatrix} = \begin{pmatrix} 0 & 1 \\ 1 & 0 \end{pmatrix}\|\psi_1\rangle$
$\|\Phi_{12}^+\rangle$	$\begin{pmatrix} -\beta \\ a \end{pmatrix} = \begin{pmatrix} 0 & -1 \\ 1 & 0 \end{pmatrix}\|\psi_1\rangle$

of the states $|\Psi_{12}^\pm\rangle$ or $|\Phi_{12}^\pm\rangle$. Bob's qubit 3 is correspondingly in one of the superposition states of $|0_3\rangle$ and $|1_3\rangle$ that are written in parentheses in Eq. (5.13). In Tab. 5.1, the left column shows the four cases resulting from Alice's measurement; the right column shows the state of Bob's photon associated with each measurement results (in vector notation).

3. *Classical communication:* From the right column of the table, we see that Bob is only one step away from realizing the desired state $|\psi_1\rangle = a\,|0\rangle + \beta\,|1\rangle$ for his qubit 3. Depending on Alice's measurement result, he only needs to perform a simple unitary transformation (the inverse of the matrices given in the table). The problem: He does not know Alice's result. Alice still has to inform him about the result of her Bell measurement. This can be done through a classical communication channel.

4. *Unitary transformation by Bob:* Based on this information, Bob performs the unitary transformation on his qubit given in the table. Qubit 3 is now in state $|\psi_1\rangle$; teleportation is complete.

ℹ Although the state of qubit 1 is exactly transferred to qubit 3, quantum teleportation does not violate the no-cloning theorem. Alice's measurement destroys both the entanglement between qubits 2 and 3 and the original state of qubit 1, and there is no way for Alice to restore it.

Also, no information is transmitted faster than the speed of light. Teleportation always requires classical communication. Without it, Bob does not know which transformation to perform on qubit 3, and teleportation does not work. No information is transmitted faster than the speed of light, and there is no conflict with the theory of relativity. The *speed of information transmission* is limited by the speed of the classical communication.

With quantum teleportation, *physically tap-proof communication* is possible. After the entangled pair is distributed to Alice and Bob, the state is directly transmitted from one qubit to the other without any way to read it without disturbance. Classical communication can be done openly because the result of the Bell measurement is useless without having the qubits.

ℹ A Bell measurement is difficult to perform because it has to be done on two photons at the same time. They have to be detected not as two individual photons but as a common two-photon state. Since they come from different sources, this places high experimental demands on the temporal, spatial, and spectral indis-

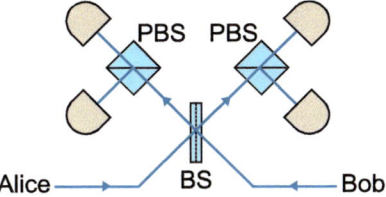

Figure 5.11: Typical scheme for performing a Bell measurement.

tinguishability of the two photons. A typical setup for a polarization Bell measurement is shown in Fig. 5.11. The two photons are brought to interference in the first beam splitter and subsequently encounter polarizing beam splitters to measure their polarization. Interference effects characteristic of two-photon interference occur (*Hong-Ou-Mandel effect*, cf. p. 112). It turns out that in such a setup (with linear components and no other degrees of freedom), only two of the four Bell states can be distinguished. Several ways have been developed to perform Bell measurements in spite of this limitation [65, 62], but it remains one of the most difficult experimental challenges in quantum communication.

5.4 Quantum repeaters and quantum networks

Secure key distribution through quantum communication uses BB84-like or entanglement-based protocols. For technical applications, the important question is: At what data rate and over what distances can these protocols be implemented? To answer this question, one has to consider the transmission channels through which the photons propagate. Surprisingly, long distances have been demonstrated experimentally: With photons in optical fibers, secure key distribution over 420 km has been achieved (at a data rate of 6.5 bit/s) [66]; with free-space transmission, entangled photon pairs could be distributed over 1200 km using the Chinese Micius satellite [67].

The maximum distance over which information can be transmitted by quantum communication is limited by the ability to coherently transmit quantum mechanical states. Surprisingly, this question does not need to be answered for each specific implementation separately, but there is a fundamental upper limit to the transmission rate in noisy channels called the *PLOB bound* [68]. It specifies the maximum number of bits that can be transmitted per channel use and has a simple mathematical form:

$$C = -\log_2(1 - \eta), \tag{5.14}$$

with $\eta = \eta_d \times 10^{-\frac{\alpha x}{10}}$, where η_d is the detector efficiency, x is the distance, and α is the attenuation rate of the signal in the medium used. Fig. 5.12 shows a logarithmic plot of the maximum transmission rate as a function of distance. The value used for α is 0.2 dB/km, which is the attenuation rate in optical fiber at 1550 nm. It can be seen that the probability of successfully transmitting a bit decreases exponentially with distance. Distances greater than a few hundred kilometers cannot be covered in this way.

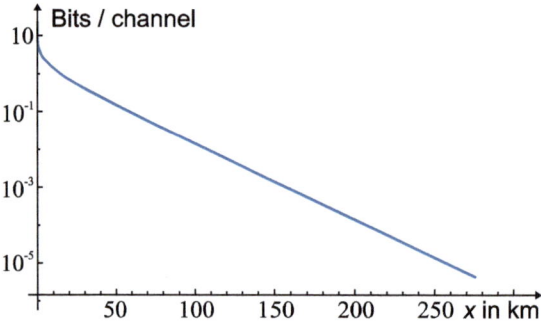

Figure 5.12: PLOB bound: Maximum transmission rate per channel as a function of distance x (absorption rate 0.2 dB/km, detector efficiency $\eta_\mathrm{d} = 1$).

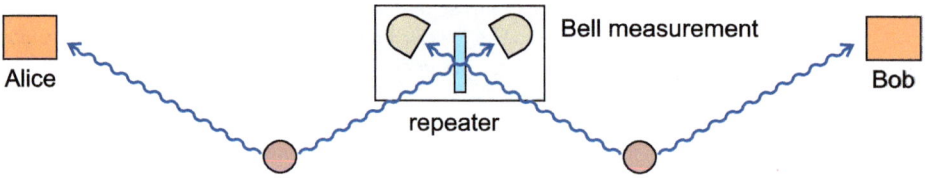

Figure 5.13: Realization of a quantum repeater by entanglement swapping.

There are similar problems with classical communication. Classical light is also exponentially attenuated in optical fibers. The solution is to refresh the signal in regular distances with *repeaters*. In practice, this is done, for example, with erbium-doped fiber amplifiers, which operate in a similar way as a laser. In quantum communication, this kind of "refreshing" is not easily possible. The no-cloning theorem prevents photons from being simply copied. An amplification mechanism for quantum states, therefore, has to be implemented differently.

Entanglement swapping and quantum repeaters

The basic idea of a *quantum repeater* [69] is illustrated in Fig. 5.13. The distance between Alice and Bob, which is too large to be covered by an entangled photon pair with any appreciable efficiency, is split into two equal parts. An entangled photon pair is distributed over each of the parts. A Bell measurement is made where the two middle photons meet. As in quantum teleportation, classical communication and local operations are required to ensure that the two outer photons are subsequently entangled. Thus, by using two pairs of entangled photons, entanglement can be extended to a range twice as large as before. This process is called *entanglement swapping*. In principle, it can be extended to any number of segments so that entanglement can, in principle, be transmitted over large distances.

Even if the principle of a quantum repeater may appear enticingly simple in Fig. 5.13 – in practice, it will not work that way. We have already seen on p. 131 that

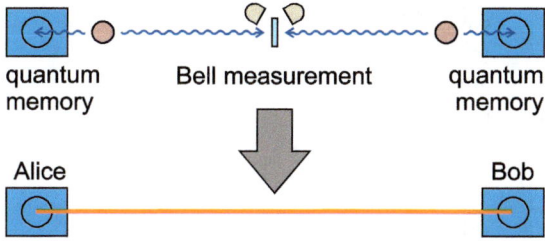

Figure 5.14: Entanglement of two quantum memories.

in a Bell measurement, the two photons arriving at the beam splitter have been indistinguishable – especially with respect to the simultaneity of their arrival. Over longer distances and multiple segments, this requirement is practically impossible to realize, especially since the whole thing is based on probabilistic processes. A "buffer" is needed for the entanglement. To realize this buffer, the entanglement has to be transferred from "flying qubits" – the photons needed to bridge long distances – to stationary qubits, which are able to store the state for a short time (at least a few milliseconds). This is the idea behind the concept of quantum memory.

Quantum memory

In *quantum memory*, one of the photons of an entangled photon pair is absorbed by a stationary quantum system (like an ion in a trap) so that the entanglement is preserved. The stationary quantum system (symbolized by a circle in Fig. 5.14) is now entangled with the second photon for further use. If there are two such setups whose two propagating photons are brought together in a Bell measurement, the two quantum memories become entangled (orange line in Fig. 5.14).

The result of this process, also called *entanglement distribution*, is two entangled quantum memories that may be a considerable distance apart, which, however, is still limited by the PLOB bound. To bridge larger distances, several of these arrangements are combined (Fig. 5.15 top). A number of quantum memory pairs are entangled in the manner described above. Now a Bell measurement is performed on two adjacent quantum memories that belong to different pairs (middle row in Fig. 5.15). By means of entanglement swapping, the outer quantum memories will be entangled afterwards (Fig. 5.15 bottom).

Although the experimental and technological challenges for the construction of quantum memories are large and partly unsolved, their use is considered necessary to realize quantum repeaters. The advantage over the simple scheme in Fig. 5.13 lies in the much more moderate timing requirements. Two quantum memories at a time can be entangled independently, and only then are the Bell measurements performed, extending entanglement over longer distances.

There are many options for the physical realization of quantum memories, and a number of approaches are being studied experimentally. These include single atoms or

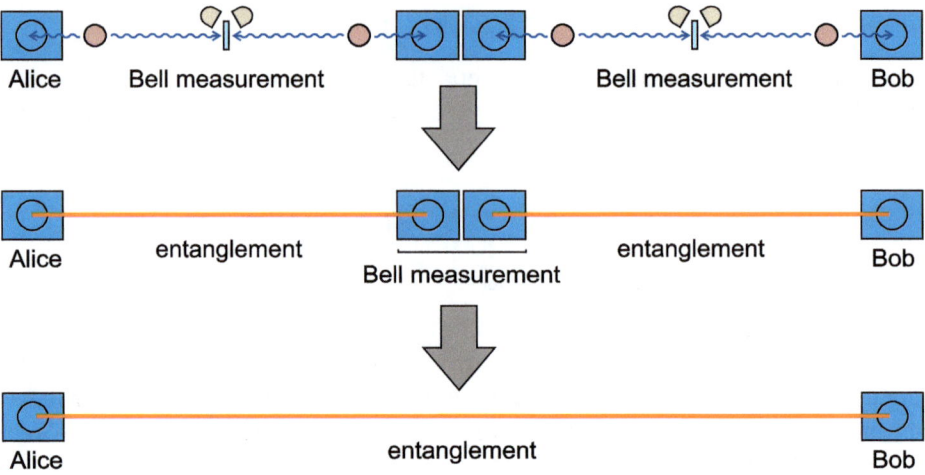

Figure 5.15: Entanglement of two distant quantum memories.

ions in traps, cold quantum gases, NV centers in diamond, or doped solid crystals. In 2022, using trapped Rydberg atoms, it was possible to entangle two quantum memories according to the scheme of Fig. 5.14 via a 33 km long optical fiber link [70]. At the University of Innsbruck, a repeater node has been realized in 2023. It consists of an ion-based quantum memory on which gate operations could be performed, coupled to 25 km of optical fiber on each side. The arrangement allowed the entanglement of photons at telecom frequencies over a distance of 50 km [71].

Entanglement purification

In the previous considerations on quantum teleportation and quantum repeaters, we have assumed that entanglement can be established between the respective partners in an ideal way. In reality, this is most often not the case. Preparation of Bell states and Bell measurements are usually not perfect, and there is noise and loss in transmission and storage. Limited detector efficiencies also contribute to imperfect entanglement of the various components – in other words, decoherence occurs.

The concept of *entanglement purification* is intended to solve this problem. If decoherence cannot be prevented, it should at least be reversed in some way. This is not possible for a single system, but if you have several entangled systems, all subject to the same decoherence mechanisms, you can sacrifice some of the systems to get a better degree of entanglement for the remaining ones. The entanglement is distilled from many noisy systems and condensed in a "purified" form onto a few remaining systems (Fig. 5.16). This is done by measurements on pairs of entangled states, local operations, and classical communication [72].

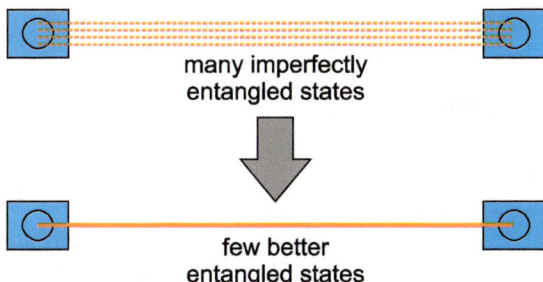

Figure 5.16: Entanglement purification: from many imperfectly entangled states, fewer better-entangled states are "distilled".

Quantum networks and the Quantum Internet

Once the components described above, which allow coherent transmission and storage of quantum information and entanglement, will reach the stage of technical application, they can be used to build networks. The focus will be on different use cases:

1. *Networks of quantum sensors*, for example, of atomic clocks, will allow ultra-precise synchronization and lead to significantly increased measurement accuracy, for example, in metrology, geodesy, or the detection of gravitational waves.
2. *Quantum communication networks*, which not only connect two partners but form a network with a large number of nodes, open up the possibility of secure network communication.
3. *Distributed quantum computing*, the interconnection of quantum computers either among themselves or with classical systems, leads to expanded possibilities and more flexibility in using quantum computers.

The set of these highly interconnected quantum communication applications is also referred to as the *quantum internet*. Building such a network is a technological challenge. Different network architectures will be used, and transitions between classical and quantum systems and between different physical realizations of quantum communication components will have to be established. Different technologies will be used in parallel and complementary ways to transmit quantum information – for example, fiber-based and satellite-based systems. Although much work remains to be done, quantum communication is one of the fastest-developing areas of quantum technology.

6 Quantum computing and quantum algorithms

6.1 Basic principles of quantum computing

Description levels for classical computing

The operation of classical computers is based on the fact that they store information in *bits*, which can take exactly two states (0 and 1). By performing logical operations on the information stored in this way, the computer moves from input to output, i. e., to the desired result. How this happens in detail can be described at different levels of hierarchy. To better understand the principle of quantum computing, it is worthwhile to briefly consider these different levels of computer architecture and then the differences between classical and quantum computing:

1. *Physical level:* Here we are concerned with the physical realization of bits by electronic devices that have two stable states, realized, for example, by high or low resistance, charge storage, or magnetic orientation. One of the two states encodes 0, the other encodes 1.

2. *Elementary logic operations with gates:* At this level, we ignore the physical realization and consider only the abstract bit states 0 and 1. They can be used to perform elementary logic operations implemented by *gates* (e. g., NOT, AND, OR). For example, an AND gate has two logic inputs and one logic output. It returns the output state 1 if both inputs have the value 1, and 0 otherwise. Simple circuits, such as adding binary numbers, can be realized by combining a few gates.

3. *Machine language and assembler:* Each computer processor specifies in its machine language sets of instructions for more complex operations, consisting of more or less extensive combinations of gate operations. These instructions are given abbreviations (such as "mov" for moving data between different memory areas) to make them more manageable for humans. In assembly language programs, these instructions are used for machine-oriented programming.

4. *Higher-level programming languages:* Even more abstract commands (such as for-loops or if-else branches) are used in higher-level programming languages like C++, Java, or Python to write application software.

5. *Application software:* This is the layer that users interact with, such as office software or games. Application software can have macro languages for programming.

Classical gates

Elementary logic operations on classical bits are described by logic gates. The most important gates for classical bits are shown in Fig. 6.1. The tables at the bottom of the figure are called *truth tables*. They indicate how the output of the gate is related to the input. For any combination of input bits, the value of the output bit can be read from the tables.

For a single classical bit, there are not many possibilities: You can leave it unchanged or invert it by turning a 0 into a 1 and vice versa. This operation is described by the NOT

https://doi.org/10.1515/9783110717457-006

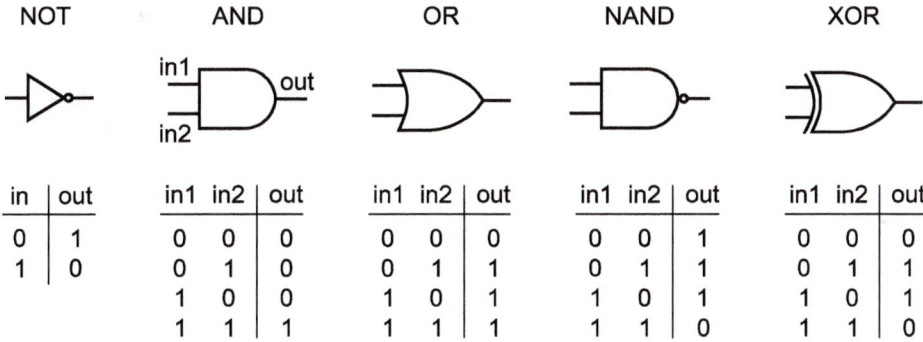

NOT		AND			OR			NAND			XOR		
in	out	in1	in2	out	in1	in2	out	in1	in2	out	in1	in2	out
0	1	0	0	0	0	0	0	0	0	1	0	0	0
1	0	0	1	0	0	1	1	0	1	1	0	1	1
		1	0	0	1	0	1	1	0	1	1	0	1
		1	1	1	1	1	1	1	1	0	1	1	0

Figure 6.1: The most important classical logic gates with their truth tables.

gate (left in Fig. 6.1). In the truth table of the NOT gate, each input bit is assigned to the inverse output bit.

The NOT gate is the only nontrivial gate for single classical bits. We will see that this is different for qubits. There are quite a few one-bit gates for qubits because they can exist in quantum mechanical superposition states and offer correspondingly richer manipulation possibilities. This is the first indication of the potentially greater power of quantum computers.

The most important classical gates with two inputs and one output are shown on the right in Fig. 6.1:

1. *AND gate:* The output is 1 if both input bits are 1, that is, if in1 = 1 *and* in2 = 1. Otherwise, the output bit has the value 0.
2. *NAND gate:* This is the inverted AND gate, i. e., AND followed by NOT.
3. *OR gate:* The output is 1 if in1 = 1 *or* in2 = 1 (especially if both bits are 1).
4. *XOR gate:* This gate performs the "exclusive OR" operation. The output bit is set to 1 if exactly one of the two input bits is 1, but not both.

In everyday language, the word "or" is sometimes used in the sense of XOR ("tea or coffee?"), but sometimes in the sense of OR ("milk or sugar?").

The NAND gate is an example of an *universal classical gate*: By combining NAND gates, it is possible to emulate any other classical gate. This is of great technological importance because any conceivable logical operation can thus be realized by implementing a single gate type and appropriate interconnections. **i**

Basic operation of a quantum computer

The basic building block of a quantum computer is the qubit. Information is stored in qubits and processed during the execution of algorithms. At the end, a result is returned. The use of quantum mechanics brings additional possibilities but also limitations. The main differences to the classical computer are:

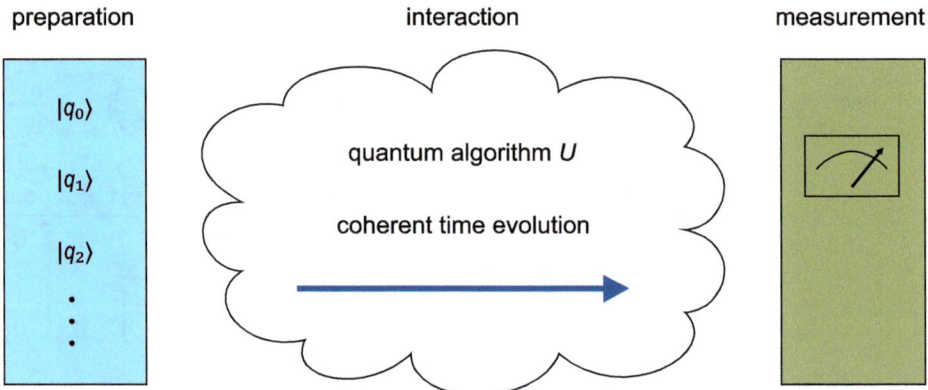

Figure 6.2: Working principle of a quantum computer in the scheme preparation – interaction – measurement.

- the *quantum parallelism* opened up by the possibility of putting qubits into arbitrary superposition states of $|0\rangle$ and $|1\rangle$, and the associated possibility of using quantum mechanical interference processes;
- the need to read out the result at the end of a calculation by a quantum measurement. This entails all the peculiarities of the quantum measurement process, particularly its statistical nature.

At the most general level, the operation of a quantum computer is described by the scheme "preparation – interaction – measurement" known from Chapter 3, which generally characterizes all controlled processes in quantum mechanics (Fig. 6.2). The system under consideration consists of a number of qubits $q_0 \ldots q_n$, whose states we denote by $|q_0\rangle \ldots |q_n\rangle$. Multiple qubits are sometimes grouped into *quantum registers*.

To enable the controlled operation of the quantum computer, the qubits must be prepared in a specific initial state. Usually, all qubits are prepared in the state $|0\rangle$ at the beginning of a computation. Since this state does not contain any information, some qubits must be provided with input. For example, when searching a database (as in the Grover algorithm considered later), the database must be encoded in the qubits. In the scheme of Fig. 6.2, this completes the preparation phase.

A *quantum algorithm* now runs on the prepared qubits. Symbolically, the algorithm is described by quantum gates, physically by microwave or laser pulses, and mathematically by a unitary transformation U on the qubits, which describes the time evolution (Fig. 6.2). The benefits of quantum computing are already becoming apparent here: For the basis states $|0\rangle$ and $|1\rangle$ the effect of the quantum algorithm is:

$$|0\rangle \to U\,|0\rangle \quad \text{and} \quad |1\rangle \to U\,|1\rangle. \tag{6.1}$$

If we now consider a qubit in a superposition state

$$|q_k\rangle = \alpha\,|0\rangle + \beta\,|1\rangle\,, \qquad\qquad (6.2)$$

the effect of the quantum algorithm U on this state is:

$$U\,|q_k\rangle = \alpha U\,|0\rangle + \beta U\,|1\rangle\,. \qquad\qquad (6.3)$$

We see that this state contains both $U\,|0\rangle$ and $U\,|1\rangle$. Thus, for qubits in a superposition state, the final state contains the result for *both* $|0\rangle$ and $|1\rangle$ basis states. This feature is called *quantum parallelism*.

Quantum parallelism: Quantum mechanical superposition states can be used to achieve that the final state of a qubit system contains the effect of a quantum algorithm on all possible basis states simultaneously.

However, there is an immediate drawback to quantum parallelism: It is *not* possible to perform parallel computations with arbitrary algorithms. The qubits have to be read out by a quantum measurement process (right in Fig. 6.2). Here, all the peculiarities of quantum measurement come in. Although the state (6.3) contains both calculation results simultaneously – they cannot be read separately with a quantum measurement. Only *one* measurement with *one* result is possible on the system. The superposition is destroyed by the measurement.

While we have not yet specified any details of the quantum algorithm, we can be sure that the laws of quantum measurement will apply to quantum computers. Thus, in a single measurement, we will get either 0 or 1 – each with a certain probability due to the statistical nature of the quantum measurement process. To take advantage of quantum parallelism, quantum algorithms need to be designed in such a way that the desired result can be extracted from this single measurement. They typically use interference between different qubit states to do this.

The role of measurement in quantum computing: A quantum algorithm must be designed so that the desired solution to the problem can be obtained as the result of a quantum measurement on the qubits.

The principle of a quantum computer can be described as follows: Using superpositions of qubit states, it works in a massively parallel manner – but in the end, you have to ask only a single question to deduce the desired result. The trick is to ask that question so cleverly that the answer solves the problem at hand.

Thus, a quantum computer operates fundamentally differently from a classical parallel computer, in which many processors perform different calculations independently of each other. Some viable ways to exploit quantum parallelism under these constraints have been found:

– If there are only a few components of the state which are of interest (as in a database search), one can try within the quantum algorithm to amplify the probability of getting exactly one of these components in the final measurement. This approach is

called *amplitude amplification*. It is used, for example, in the Grover database search algorithm (cf. Section 6.3).
–   One can use global properties of the final state, like searching for periodicities with the quantum Fourier transform. Such an approach is the basis of Shor's algorithm for factorizing large numbers (cf. Section 6.6).

Reversibility and decoherence

In order to use quantum parallelism, interference has to occur. This is crucial for successful quantum computing. Therefore, it is necessary to create and preserve superpositions between the states of a qubit. Even creating entangled states between several qubits is necessary for most quantum algorithms.

The need for interference immediately leads to an essential requirement for the design of a quantum computer, which is as simple to formulate as it is difficult to realize in practice: the system must be kept coherent in its essential parts. It is necessary to prevent the occurrence of decoherence, which destroys the ability to interfere. It is this requirement that makes it so difficult to build a quantum computer. If we recall the results of Section 3.10 about decoherence and its ubiquity, we can estimate the difficulties that must be overcome to build a quantum computer. On the whole, the following conflict arises:
–   On the one hand, the qubits must interact with each other (otherwise, no meaningful computation is possible). They also have to be manipulated individually from the outside (otherwise, no preparation, i. e., no input, is possible).
–   On the other hand, the qubits must be isolated from their environment to such an extent that the superposition states, which are essential for computation, are not subject to decoherence. Basic rule 4 (cf. p. 46) provides a criterion for this requirement: The ability to interfere is destroyed by any interaction that carries information about the state of the qubits to the outside.

Therefore, for the physical realization of qubits, systems are considered whose relevant degrees of freedom can be well isolated from the interaction with the environment but which must still be individually addressable.

Requirements for quantum gates

The need for coherence also affects the architecture of the next higher level of description: the level of logic operations and quantum gates. In fact, keeping the system coherent means that irreversibility must not occur. Irreversibility is an indication of loss of coherence. A coherent quantum process is characterized by being reversible in time. It can run in the forward direction just as well as in the reverse direction. Mathematically speaking, all operations performed on qubits must be unitary. The action of quantum gates is mathematically described by unitary transformations on the state of the qubits.

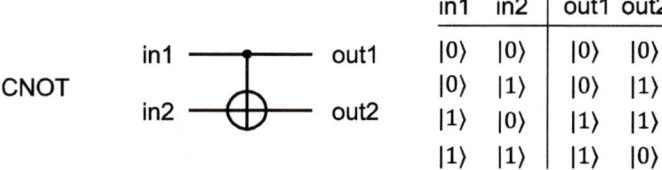

		in1	in2	out1	out2
CNOT	in1 ——— out1 in2 ——— out2	$\|0\rangle$	$\|0\rangle$	$\|0\rangle$	$\|0\rangle$
		$\|0\rangle$	$\|1\rangle$	$\|0\rangle$	$\|1\rangle$
		$\|1\rangle$	$\|0\rangle$	$\|1\rangle$	$\|1\rangle$
		$\|1\rangle$	$\|1\rangle$	$\|1\rangle$	$\|0\rangle$

Figure 6.3: The CNOT gate is reversible: the final state can be uniquely deduced from the initial state.

> To avoid loss of coherence, quantum algorithms have to be *reversible*. All operations have to be described mathematically as unitary transformations on the qubits.

For quantum computing, this means that information must not be destroyed during computation. For example, in quantum algorithms, qubits cannot simply be erased (i. e., set to $|0\rangle$ regardless of their previous state). If this happens, the reverse process would not be possible because the initial state cannot be reconstructed from the state $|0\rangle$. Therefore, in order to return qubits to the initially prepared state $|0\rangle$ without erasing them, quantum algorithms regularly run parts of the code backwards. This process is called *uncomputation*.

Only certain types of gates are suitable for coherent computation. For example, an OR gate is not reversible: if the final state at the output is 1, the initial state cannot be inferred: it could have been 01, 10, or 11. The OR operation is not reversible; it is a one-way street for information.

The same is generally true for gates with two inputs and one output (such as AND or NAND). None of the classical two-bit gates shown in Fig. 6.1 are reversible. To avoid loss of information, quantum gates must always have as many outputs as inputs. Therefore, instead of AND and OR gates, quantum computers use reversible gates such as the CNOT (Controlled-NOT) gate shown in Fig. 6.3. It has two inputs and two outputs that uniquely map initial and final states. Mathematically, reversible quantum gates are described by unitary transformations of the qubit states.

DiVincenzo criteria

As we have seen, building a quantum computer is not easy. In fact, for a long time, it was not at all clear that it would ever be possible to build useful quantum computers. The requirements for coherence and controllability are enormous. In 2000, David DiVincenzo formulated a list of criteria that need to be met for the successful implementation of a quantum computer [73]:

1. *A scalable system with well-defined qubits.* Scalable means that it must be technically possible to go beyond prototypes with a few qubits with reasonable effort. The equipment used to control and drive individual qubits needs to remain manageable when scaling to 50 or more logical qubits. Similarly, the effort required for error correction and decoherence suppression must be manageable.

2. *The ability to prepare the qubits in a well-defined initial state.* The controlled execution of quantum algorithms requires that all qubits can be prepared to a well-defined initial state at the beginning of the computation, usually the state $|0\rangle$.
3. *Long decoherence times, much longer than the timescales for gate operations.* For quantum algorithms to be successfully implemented, a sufficiently large number of gate operations need to be possible before the system loses coherence.
4. *The experimental feasibility of a universal set of quantum gates.* To realize quantum gates, it must be possible to manipulate the system with laser or microwave pulses.
5. *The ability to make measurements on individual qubits.* It must be possible to measure individual qubits to read out the computation result.

As described in Chapter 1, the current development of quantum computers relies on a number of competing approaches that attempt to realize these requirements using different physical systems, such as single ions in traps, superconducting qubits, neutral atoms, or quantum dots in semiconductors.

6.2 A simple quantum algorithm: Quantum Penny Flip

Let us now look at the first quantum algorithm. We will use a simple example to discuss the advantages of quantum computers and to introduce some quantum gates. It is not a sophisticated algorithm for a practical application but just a simple game where the use of quantum physics gives a winning advantage. It is called *Quantum Penny Flip* and was proposed in 1999 by David A. Meyer [74].

The classical rules

The starting situation is the classic coin toss with two players (Alice and Bob): A coin is tossed. If it lands on "heads", Alice wins, if "tails" comes up, Bob wins. In order to relate the game to a quantum algorithm later on, we have to slightly rewrite the rules:

1. Alice places the coin in a state of her choice (heads or tails on top) in a box that cannot be looked into. Even by touch, it is impossible to tell which side is up.
2. Bob reaches into the box and has the choice of flipping the coin or not. Alice cannot see what he is doing.
3. Alice reaches into the box and performs an action of her choice (flip or not flip).
4. The coin is uncovered and the result is read.

With these rules, we have formulated the rules of the game according to the scheme "preparation – interaction – measurement" (Fig. 6.4). Although the roles of Bob and Alice are not exactly symmetric, it is clear that there is no winning strategy for either player. The chances are evenly distributed.

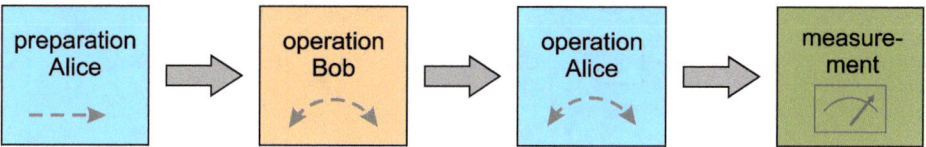

Figure 6.4: The game Quantum Penny Flip in the scheme preparation – interaction – measurement.

Quantum version of the game

The quantum version of the algorithm can now be formulated in analogy to the rules just established. The classical coin is replaced by a qubit, which can be in any superposition of $|0\rangle$ and $|1\rangle$. Another difference is that Alice knows the laws of quantum physics and can perform all possible operations on the qubit. Bob is limited to the classical operations of flipping or not flipping.

For the sake of illustration, we will describe the algorithm both in Dirac and matrix notation. For more complex examples, matrix notation quickly becomes cumbersome. Therefore, the Dirac notation is usually preferred.

We are dealing with a single qubit. Its basis states are column vectors with two components:

$$|0\rangle = \begin{pmatrix} 1 \\ 0 \end{pmatrix} \quad \text{``heads''},$$

$$|1\rangle = \begin{pmatrix} 0 \\ 1 \end{pmatrix} \quad \text{``tails''}.$$

As usual, we assume that the qubit is initially prepared in the state $|0\rangle$.

Qubit operations

Since only one qubit is involved, the algorithm is limited to *one-bit operations*. Classically, these operations are very limited: they correspond to Bob's action options flip (NOT gate) and non-flip (identity $\mathbb{1}$).

Quantum mechanically, there are many more operations possible on a qubit. As a reminder, the states $|0\rangle$ and $|1\rangle$ are represented by the north and south poles of the Bloch sphere. In between there is a continuum of possible states (the whole surface of the Bloch sphere). Correspondingly, there is a large number of one-qubit operations to get from one state to another. The table on p. 205 gives an overview of the most important quantum gates. For the algorithm considered here, the gates shown in Fig. 6.5 are relevant. Besides the identity operation $\mathbb{1}$, which does not change the state, we consider the quantum mechanical NOT gate, also called *Pauli-X*, and the *Hadamard gate*.

Pauli-*X* gate

The Pauli-*X* gate transforms the state $|0\rangle$ into $|1\rangle$ and conversely $|1\rangle$ into $|0\rangle$. It describes the flip of the coin. In matrix notation, it can be represented by a 2×2 matrix (it is one

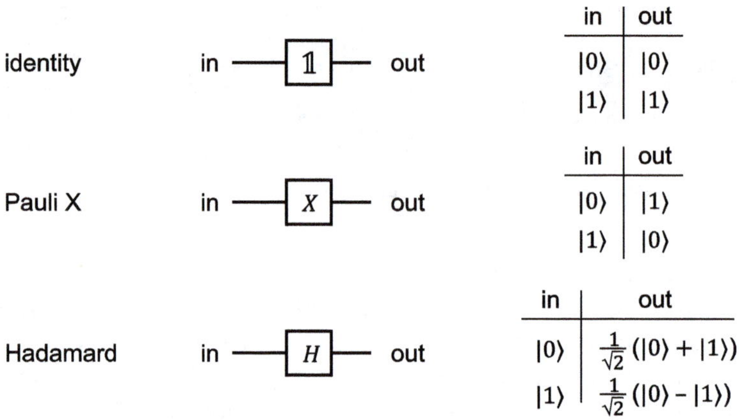

			in	out
identity	in —[$\mathbb{1}$]— out		$\lvert 0 \rangle$	$\lvert 0 \rangle$
			$\lvert 1 \rangle$	$\lvert 1 \rangle$

			in	out
Pauli X	in —[X]— out		$\lvert 0 \rangle$	$\lvert 1 \rangle$
			$\lvert 1 \rangle$	$\lvert 0 \rangle$

			in	out
Hadamard	in —[H]— out		$\lvert 0 \rangle$	$\frac{1}{\sqrt{2}}(\lvert 0 \rangle + \lvert 1 \rangle)$
			$\lvert 1 \rangle$	$\frac{1}{\sqrt{2}}(\lvert 0 \rangle - \lvert 1 \rangle)$

Figure 6.5: The quantum gates required for the Quantum Penny Flip algorithm.

of the Pauli matrices of p. 68):

$$X = \begin{pmatrix} 0 & 1 \\ 1 & 0 \end{pmatrix}. \tag{6.4}$$

By explicit calculation, it can be confirmed that this matrix acts on the basis states as desired:

$$X\lvert 0 \rangle = \begin{pmatrix} 0 & 1 \\ 1 & 0 \end{pmatrix}\begin{pmatrix} 1 \\ 0 \end{pmatrix} = \begin{pmatrix} 0 \\ 1 \end{pmatrix} = \lvert 1 \rangle \quad \text{and} \quad X\lvert 1 \rangle = \begin{pmatrix} 0 & 1 \\ 1 & 0 \end{pmatrix}\begin{pmatrix} 0 \\ 1 \end{pmatrix} = \begin{pmatrix} 1 \\ 0 \end{pmatrix} = \lvert 0 \rangle.$$

Unlike the classical NOT gate, the Pauli-X gate is also defined for quantum mechanical superposition states. The result follows directly from the linearity of the operation. If $\lvert \psi_{in} \rangle = \alpha \lvert 0 \rangle + \beta \lvert 1 \rangle$, then:

$$\lvert \psi_{out} \rangle = X \lvert \psi_{in} \rangle = \alpha \lvert 1 \rangle + \beta \lvert 0 \rangle. \tag{6.5}$$

In Dirac notation, the Pauli-X gate can be expressed as follows (cf. p. 63):

$$X = \lvert 0 \rangle \langle 1 \rvert + \lvert 1 \rangle \langle 0 \rvert. \tag{6.6}$$

Hadamard gate
The Hadamard gate H transforms the basis states $\lvert 0 \rangle$ and $\lvert 1 \rangle$ into superposition states:

$$H\lvert 0 \rangle = \frac{1}{\sqrt{2}}(\lvert 0 \rangle + \lvert 1 \rangle) \quad \text{and} \quad H\lvert 1 \rangle = \frac{1}{\sqrt{2}}(\lvert 0 \rangle - \lvert 1 \rangle). \tag{6.7}$$

It is one of the most commonly used gates in quantum algorithms because without superposition there is no interference, and without interference, there is no quantum ad-

a) Bob flips

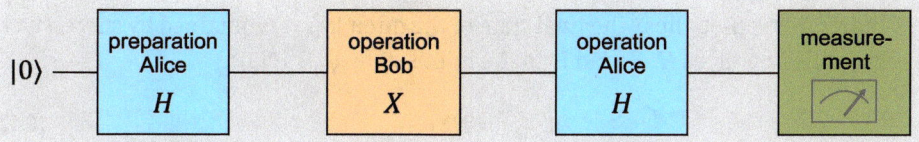

b) Bob does not flip

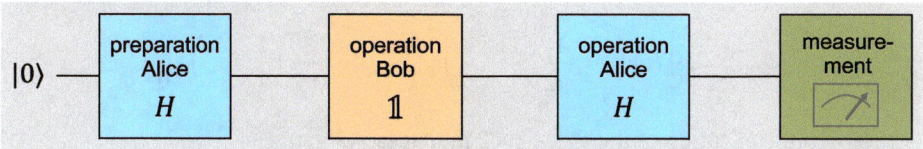

Figure 6.6: The sequence of possible qubit operations in Quantum Penny Flip.

vantage. Therefore, it is often used at the very beginning of a quantum algorithm to create superposition states.

In matrix notation, the Hadamard gate can be written as follows

$$H = \frac{1}{\sqrt{2}} \begin{pmatrix} 1 & 1 \\ 1 & -1 \end{pmatrix}. \tag{6.8}$$

Again, we demonstrate the effect on the basis states by explicit calculation:

$$H \left| 0 \right\rangle = \frac{1}{\sqrt{2}} \begin{pmatrix} 1 & 1 \\ 1 & -1 \end{pmatrix} \begin{pmatrix} 1 \\ 0 \end{pmatrix} = \frac{1}{\sqrt{2}} \begin{pmatrix} 1 \\ 1 \end{pmatrix} = \frac{1}{\sqrt{2}} (\left| 0 \right\rangle + \left| 1 \right\rangle), \tag{6.9}$$

$$H \left| 1 \right\rangle = \frac{1}{\sqrt{2}} \begin{pmatrix} 1 & 1 \\ 1 & -1 \end{pmatrix} \begin{pmatrix} 0 \\ 1 \end{pmatrix} = \frac{1}{\sqrt{2}} \begin{pmatrix} 1 \\ -1 \end{pmatrix} = \frac{1}{\sqrt{2}} (\left| 0 \right\rangle - \left| 1 \right\rangle). \tag{6.10}$$

In Dirac notation, the expression for the Hadamard gate is as follows:

$$H = \frac{1}{\sqrt{2}} [\left| 0 \right\rangle \left\langle 0 \right| + \left| 1 \right\rangle \left\langle 0 \right| + \left| 0 \right\rangle \left\langle 1 \right| - \left| 1 \right\rangle \left\langle 1 \right|]. \tag{6.11}$$

The winning strategy in Quantum Penny Flip

Alice's winning strategy in Quantum Penny Flip is very simple: whenever it is her turn, she performs the operation H on the coin. If she follows this strategy, she will win no matter what Bob does.

Fig. 6.6 shows the sequence of possible operations on the qubit. At the beginning, the qubit is prepared in the state $\left| 0 \right\rangle$. First, it is Alice's turn, then Bob's turn, then Alice's turn again. If Alice always uses H, there are two possibilities for the progress of the game: Bob can flip the coin (operation X) or not (operation $\mathbb{1}$). Using the explicit expressions for the quantum gates given earlier, we can determine the outcome in both cases.

Case (a): Bob flips

For the first case (Bob flips), the final state of the qubit $|\psi_{\text{out}}\rangle$ is obtained by successively applying the operators H, X, and H to the initial state $|\psi_{\text{in}}\rangle = |0\rangle$:

$$|\psi_{\text{out}}\rangle = HXH\,|\psi_{\text{in}}\rangle\,. \tag{6.12}$$

In matrix notation, we write explicitly:

$$|\psi_{\text{out}}\rangle = HXH\,|0\rangle = \frac{1}{\sqrt{2}}\begin{pmatrix} 1 & 1 \\ 1 & -1 \end{pmatrix}\begin{pmatrix} 0 & 1 \\ 1 & 0 \end{pmatrix} \times \frac{1}{\sqrt{2}}\begin{pmatrix} 1 & 1 \\ 1 & -1 \end{pmatrix}\begin{pmatrix} 1 \\ 0 \end{pmatrix}. \tag{6.13}$$

By matrix multiplication of the three 2×2 matrices, we get:

$$|\psi_{\text{out}}\rangle = \begin{pmatrix} 1 & 0 \\ 0 & -1 \end{pmatrix}\begin{pmatrix} 1 \\ 0 \end{pmatrix} = \begin{pmatrix} 1 \\ 0 \end{pmatrix} = |0\rangle\,. \tag{6.14}$$

So, the result is $|0\rangle$ = heads, and Alice has won.

Case (b): Bob does not flip

To compare the two formalisms, we approach the second possibility (Bob does not flip) in Dirac notation. Since the unit operator $\mathbb{1} = |0\rangle\langle 0| + |1\rangle\langle 1|$ represents doing nothing, its presence actually has no effect on the final result. However, we will keep it to practice using bras and kets:

$$|\psi_{\text{out}}\rangle = H\mathbb{1}H\,|0\rangle = \frac{1}{\sqrt{2}}\big[|0\rangle\langle 0| + |1\rangle\langle 0| + |0\rangle\langle 1| - |1\rangle\langle 1|\big] \times \big[|0\rangle\langle 0| + |1\rangle\langle 1|\big]$$
$$\times \frac{1}{\sqrt{2}}\big[|0\rangle\langle 0| + |1\rangle\langle 0| + |0\rangle\langle 1| - |1\rangle\langle 1|\big]\,|0\rangle\,. \tag{6.15}$$

The multiplication of the terms becomes easier if we take into account $\langle 0|1\rangle = 0$ and do not write down the corresponding terms from the outset. In this way, we obtain quite easily:

$$|\psi_{\text{out}}\rangle = \big[|0\rangle\langle 0| + |1\rangle\langle 1|\big]\,|0\rangle = |0\rangle\,. \tag{6.16}$$

Again the result is heads, and again Alice has won.

Overall, we have found a safe winning strategy for Alice. No matter what Bob does – flip or not flip: as long as Alice uses the Hadamard gate on both moves, she cannot lose. This is a striking difference from the classical game, where there is no winning strategy for either player. The use of superposition states is crucial to Alice's success, and in particular, with the Hadamard gate, it is a genuine quantum gate that forms the basis of her winning strategy.

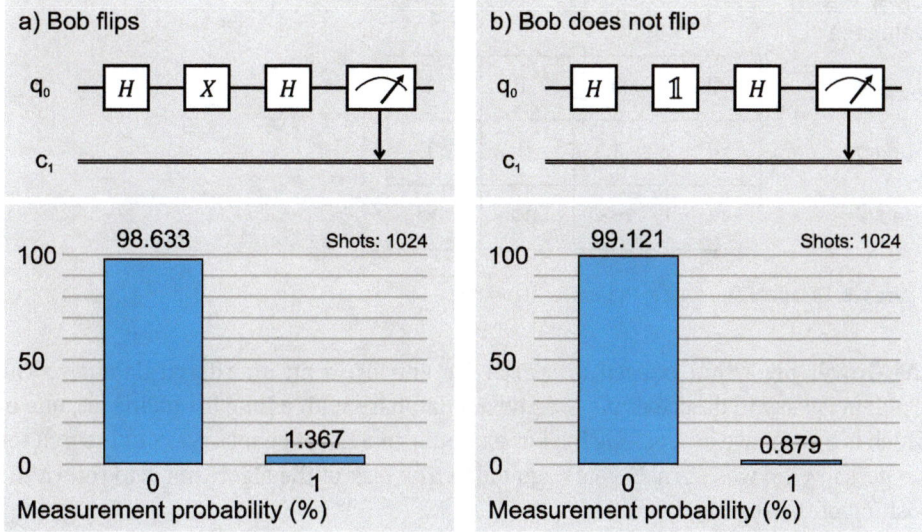

Figure 6.7: Result of a calculation on a quantum computer. The classical bit c_1 contains at the end the result of the quantum measurement on the qubit q_0.

Several quantum computer manufacturers offer free cloud-based access to their systems to interested users. We can try the Quantum Penny Flip algorithm on such a quantum computer. Programming is done either graphically with symbols for the quantum gates or in one of the programming languages developed for quantum algorithms.

    The symbolic representation of the programming for the two cases is shown in Fig. 6.7 above; the computational result is shown below. Due to decoherence, the results are noisy. Therefore, 1024 runs are performed for each of the two cases. The result confirms the winning strategy for Alice: except for the noise, the measurement at the qubit q_0 yields the output 0 (= heads) in all cases. The classical bit c_1 is used to store the result of the measurement.

6.3 Quantum database search with the Grover algorithm

The Grover algorithm [75] is one of the most prominent quantum algorithms because – unlike the quantum penny flip – it solves an important practical problem while providing a significant speedup over classical algorithms. We will discuss the Grover algorithm in detail, using this example to discuss quantum gate operations and quantum computer programming. The Grover algorithm is simple enough to understand with moderate effort but complex enough to use it to discuss many important aspects of quantum computing. A wealth of additional information on the most important algorithms in quantum computing, especially from a mathematical point of view, can be found in the standard work by Nielsen and Chuang [76].

No. in the database	1	2	3	4
qubit 1	0	0	1	1
qubit 2	0	1	0	1
oracle	no	yes	no	no

Figure 6.8: Database structure in the Grover algorithm.

The Grover algorithm is used to search for entries in an unordered database. The problem is easy to describe: We are given a database with a long list of entries, one of which is the one we are looking for. For example, in a list of people, we could search for the person who was born 25 years ago today. The task of the algorithm is to return the position of the searched entry.

Oracle

In reality, however, a quantum algorithm is unlikely to search a table of fixed entries. Instead, it focuses on scenarios where a function f needs to be evaluated to determine which entry is being searched for. It returns "yes" if it is the entry we are looking for and "no" otherwise. This function, which must be implemented as a quantum algorithm, is called *oracle* and is treated as a "black box". This kind of abstract approach is common in computer science.

The oracle is specific to the problem at hand. Its details are not important for the functioning of the algorithm. To understand the following, we only need to state: The oracle does not find the searched state in the database. It only *recognizes* it if it receives it as input and gives a corresponding feedback.

Database structure

Due to the black-box nature of the oracle, the Grover algorithm has a minimal database structure. It consists only of the numbering of the entries and an identifier of the searched entry – the oracle's answer. In a spreadsheet, this would correspond to the row numbers plus another column containing "yes" for the searched entry and "no" for all others.

Fig. 6.8 shows the database structure for $N = 4$ entries. The numbering of the entries is done with the two upper qubits. Classically, n bits can be used to number $N = 2^n$ states. The same is true for qubits if we restrict ourselves to the basis states $|0\rangle$ and $|1\rangle$. So, with 2 qubits, we can number 4 entries, with 3 qubits, we can number 8 entries, and so on. The four states that can be distinguished with two qubits are $|00\rangle$, $|01\rangle$, $|10\rangle$, $|11\rangle$. One of them is the state we are looking for. In the figure, we assume that the state $|01\rangle$ is identified by the oracle as the state we are looking for. It is marked with "yes".

In most representations of the Grover algorithm, the implementation of the oracle is omitted, and one of the entries is arbitrarily marked "by hand". This gives the impression that the search algorithm can only find something that is already known beforehand. To avoid this difficulty, we will explicitly implement a simple oracle later.

Quantum parallelism

The Grover algorithm assumes that calling the oracle is the time-consuming part of the database search. The algorithm is fast if the number of calls to the oracle is kept to a minimum. This is where quantum parallelism comes in.

In a classical search algorithm, the search in an unordered database proceeds as follows: the algorithm works through the list of entries from the beginning to the end. For each entry, there is a call of the oracle to see if it is the searched entry. For M database entries on average $M/2$ calls of the oracle are necessary until the searched entry is found.

The Grover algorithm exploits quantum parallelism. Using superposition states, the oracle can act on all database entries *simultaneously* and mark the searched entry with a 1 in a single call. While this is quantum mechanically possible, it does not solve the problem. The algorithm has not fulfilled its task until it has also revealed the searched state in a final measurement. This is the real problem of the Grover algorithm: to obtain the state identified by the oracle with a high probability as the result of a suitable quantum measurement. To achieve this, an iterative process called *amplitude amplification* is performed. The goal is to make the state we are looking for so dominant in the quantum state of the whole system that it will be found in the measurement with a high probability.

It turns out that the number of required oracle calls does not scale with M but with $\sqrt{M}$. This gives the Grover algorithm a quadratic advantage over classical search algorithms.

Principle of the Grover algorithm

The principle of the Grover algorithm is shown in Fig. 6.9. It requires a number of qubits $q_1 \ldots q_N$, which must be large enough to number the database entries. Furthermore, one or more auxiliary or *ancilla qubits* are needed to call the oracle, which does not take part in the actual calculation.

First, all qubits are put into a superposition state by applying Hadamard gates (no superposition means no interference, and no interference means no quantum advantage). Then the same sequence follows several times in succession: calling the oracle, followed by the "Grover diffuser" – a sequence of quantum gates whose operation we will explain later. This sequence of oracle and diffuser has to be called on the order of $\sqrt{M}$ times. At the end, a measurement is made on the qubit system. Due to the action of the oracle-diffuser sequence, the probability of finding the state for which the oracle gave the answer "yes" is very high. Such a process, in which the searched state is selec-

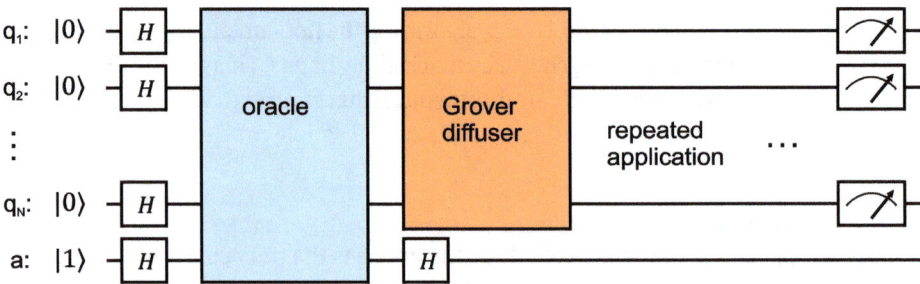

q_1: $|0\rangle$ — H —

q_2: $|0\rangle$ — H —

$\vdots$ oracle | Grover diffuser | repeated application ...

q_N: $|0\rangle$ — H —

a: $|1\rangle$ — H —— H —

Figure 6.9: Principle of the Grover algorithm. The oracle is called multiple times (order of $\sqrt{M}$ calls). The ancilla qubit, which plays a role in calling the oracle, is labeled with a.

tively amplified at the expense of all other possible states, characterizes the amplitude amplification procedure.

Simplest example: Grover algorithm with 2 qubits

It is almost impossible to understand the exact way the Grover algorithm works on the first try. Therefore, we will first consider the simplest example with $n = 2$ qubits, which can number a database with $N = 4$ states. For now, we will refrain from explicitly modeling the operation of the oracle and will only specify its effect on the state of the qubits q_1 and q_2. Therefore, we do not need an ancilla qubit yet.

1. *Step 1: Symmetrical state*

   The two qubits q_1 and q_2 are brought into a superposition state of $|0\rangle$ and $|1\rangle$ by applying the Hadamard gate (cf. p. 144). So, the initial state is

   $$|\psi_1\rangle = H_{(1)}H_{(2)}|0\rangle_1|0\rangle_2. \tag{6.17}$$

   The indices 1 and 2 at the states denote the respective qubits; the index at the operator indicates which qubit it acts on. With Eq. (6.7) we get:

   $$|\psi_1\rangle = \frac{1}{\sqrt{2}}(|0\rangle_1 + |1\rangle_1) \otimes \frac{1}{\sqrt{2}}(|0\rangle_2 + |1\rangle_2), \tag{6.18}$$

   or after multiplication and with a slightly simplified notation:

   $$|\psi_1\rangle = \frac{1}{2}[|00\rangle + |01\rangle + |10\rangle + |11\rangle]. \tag{6.19}$$

   This state is also called *symmetric state* $|s\rangle$ and will play an important role later.

2. *Step 2: Effect of the oracle*

   The effect of the oracle is to change the sign of the state we are looking for to a minus sign. In our example, it is the state $|01\rangle$. The total state after calling the oracle is thus

   $$|\psi_2\rangle = \frac{1}{2}[|00\rangle - |01\rangle + |10\rangle + |11\rangle]. \tag{6.20}$$

We can rewrite it as follows:

$$|\psi_2\rangle = |s\rangle - |01\rangle. \tag{6.21}$$

3. *Step 3: Grover diffuser*
   The Grover diffuser performs the amplitude amplification for the state that is marked by the minus sign. The diffuser is described by the operator $S = 2\,|s\rangle\,\langle s| - \mathbb{1}$:

$$
\begin{aligned}
|\psi_3\rangle &= S\,|\psi_2\rangle \\
&= \left[2\,|s\rangle\,\langle s| - \mathbb{1}\right]\left[|s\rangle - |01\rangle\right] \\
&= 2\,|s\rangle\,\underbrace{\langle s|s\rangle}_{=1} - |s\rangle - 2\,|s\rangle\,\underbrace{\langle s|01\rangle}_{=1/2} + |01\rangle,
\end{aligned}
\tag{6.22}
$$

where Eq. (6.19) was used to calculate the scalar product. All contributions containing $|s\rangle$ cancel (destructive interference), and we obtain:

$$|\psi_3\rangle = |01\rangle. \tag{6.23}$$

Already, after a single application of the sequence oracle-diffuser, the initial state $|s\rangle$ has evolved to the entry marked by the oracle, i. e., the searched state. It is found with probability 1 in the final measurement. Thus, the Grover algorithm has returned the searched database element. If the oracle had marked another entry with a minus sign, this entry would have been found.

It is not typical that the database entry can be found with only one application of the oracle-diffuser sequence. This happens only in the special case of two qubits. Similarly, it is generally not possible to identify the searched state with probability 1. In usual cases, the probability of finding it in a measurement is only significantly increased.

Geometric interpretation of the Grover algorithm

We have convinced ourselves by explicit calculation that the Grover algorithm works – at least for the case of two qubits. However, the calculation does not lead to a clear understanding of how the algorithm works. You can see that it works, but you do not understand why.

There are two common visualizations of how the Grover algorithm works. They are known as "inversion at the mean" and "rotation in a two-dimensional state space". It is interesting to note that Grover himself arrived at his algorithm by a completely different route (which he describes in [77]): Originally, he considered the qubits as a long chain of "atoms" in a continuum. The process of marking the searched qubit by the oracle creates a kind of potential well into which the state of the whole system then "diffuses" (hence the name "diffuser"). The two steps of the algorithm (oracle and diffuser) are already visible in this approach.

Today, the Grover algorithm is often illustrated by a geometric interpretation, in which the action of the algorithm is seen as a rotation in the state space. This description is so general that we can directly generalize our previous considerations to the case of n qubits.

With n qubits, the database has $N = 2^n$ states. At the beginning of the algorithm, the system is put into the symmetric state $|s\rangle$ by applying the Hadamard gate to all qubits. For n qubits, $|s\rangle$ has the following form:

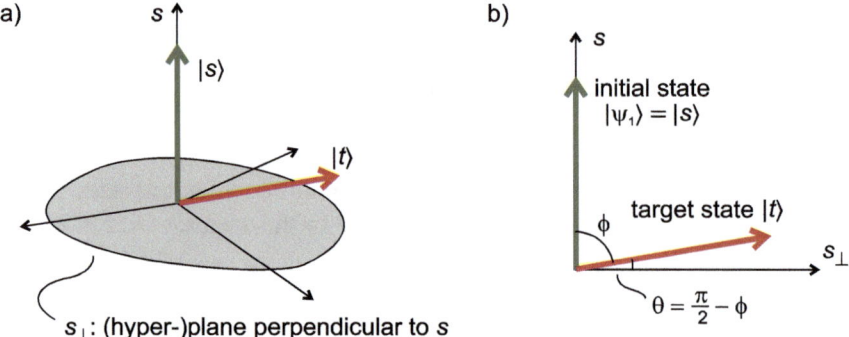

Figure 6.10: (a) Geometric representation of the states involved, (b) reduction to quasi-two-dimensional state space.

$$|s\rangle = \frac{1}{\sqrt{N}}\Big[|00\ldots0\rangle + |00\ldots1\rangle + \cdots + |11\ldots1\rangle\Big]. \tag{6.24}$$

One of the states in the square brackets is the target state, which we denote by $|t\rangle$.

The goal of amplitude amplification is to influence the initial state $|s\rangle$ by a sequence of unitary operations in such a way that it approaches the target state $|t\rangle$ as closely as possible. In the process of amplitude amplification, the state of the system is "rotated" towards the target state $|t\rangle$ so that this state is found with high probability when measured. The problem is that the target state is not known a priori and can only be marked by the oracle. To analyze the algorithm, we concede knowledge of the target state $|t\rangle$ and later examine how the algorithm deals with this difficulty.

Fig. 6.10 (a) shows the geometry of the states involved. For n qubits, the state space is 2^n-dimensional (so for 2 qubits it is four-dimensional). The coordinates are chosen in a special way: The coordinate axes *not* run in the direction of the previously considered basis vectors $|00\ldots0\rangle,\ldots,|11\ldots1\rangle$, but the coordinate system is oriented so that one axis s points in the direction of the vector $|s\rangle$. In addition, there are 2^n-1 perpendicular axes, which are only hinted at in the figure and will not play a major role in the following. For all state vectors, we will only distinguish between components parallel to $|s\rangle$ and perpendicular to it (notation $s_\perp$). This leads to the quasi-two-dimensional representation in Fig. 6.10(b).

In the high-dimensional state space, the target vector $|t\rangle$ is almost perpendicular to $|s\rangle$ at the beginning. We can already see this in our 2-qubit example: the target vector is $|10\rangle$, and from Eq. (6.19), we read that it is only one of four equal components of $|s\rangle$. In the case of n qubits, the target vector is one of the $N = 2^n$ components of $|s\rangle$. If we form the scalar product between $|s\rangle$ and $|t\rangle$, we obtain with Eq. (6.24):

$$\langle s|t\rangle = \frac{1}{\sqrt{N}} = \cos\phi. \tag{6.25}$$

Geometrically, the scalar product can be visualized by the angle ϕ between the vectors $|s\rangle$ and $|t\rangle$ (Fig. 6.10(b)).

As already mentioned, the aim of the Grover algorithm is to rotate the initial state $|s\rangle$ to the target state $|t\rangle$, i. e., by the angle ϕ. This can be achieved only indirectly and in several steps. One has to simulate a rotation by a sequence of two axis reflections. These axis reflections are produced by the oracle and the Grover diffuser.

(a) *Effect of the oracle as an axis reflection of the initial state $|\psi_1\rangle$ on the vector $|t\rangle$*

The oracle inverts the sign for the component of the initial state $|\psi_1\rangle$ that is in the direction of the target vector $|t\rangle$. Mathematically, this can be described by the operation $-T = \mathbb{1} - 2|t\rangle\langle t|$ acting on the initial state:

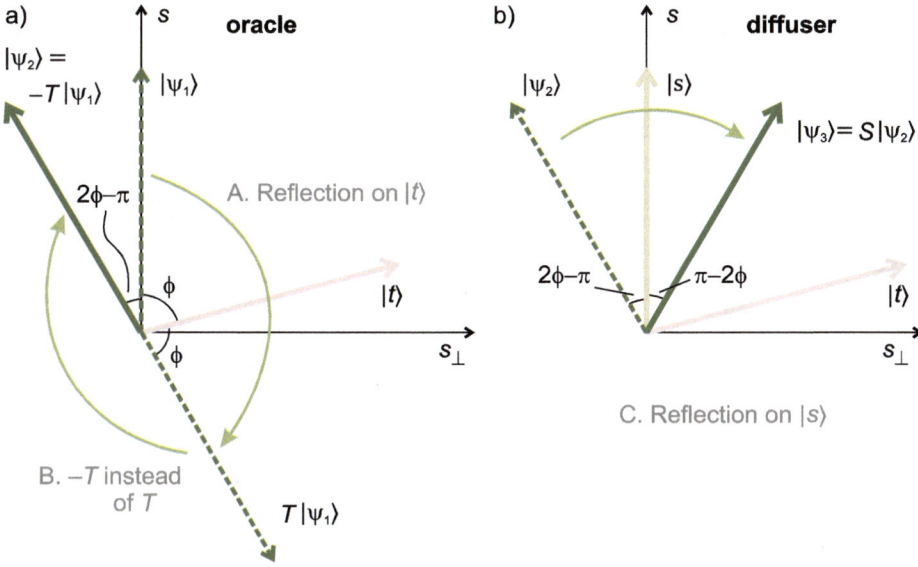

Figure 6.11: Effect of the oracle and the Grover diffuser.

$$-T = \underbrace{-|t\rangle\langle t|}_{\substack{\text{Projection on }|t\rangle \\ \text{gets minus sign.}}} + \underbrace{\left(\mathbb{1} - |t\rangle\langle t|\right)}_{\substack{\text{Orthogonal components} \\ \text{remain unchanged.}}} . \tag{6.26}$$

The negative of this, i. e., the operation $T = 2|t\rangle\langle t| - \mathbb{1}$, corresponds geometrically to a reflection of the initial state $|\psi_1\rangle$ on the $|t\rangle$ axis. In an axis reflection, the orthogonal components get a minus sign while the parallel component remains unchanged. The effect of the oracle $-T$ is illustrated in Fig. 6.11(a). By examining the angles, one can see that the state after the application of the oracle, $|\psi_2\rangle = -T|\psi_1\rangle$, has an angle of $2\phi - \pi$ with respect to the s-axis.

(b)   *Effect of the diffuser as an axis reflection on the vector $|s\rangle$*
The mathematical formula for the Grover diffuser has already been given in Eq. (6.22):

$$S = 2|s\rangle\langle s| - \mathbb{1}. \tag{6.27}$$

It has the same form as T, except that $|t\rangle$ is replaced by $|s\rangle$. This means that geometrically the Grover diffuser is an axis reflection on the vector $|s\rangle$. The effect is illustrated in Fig. 6.11(b). The angle of the resulting state $|\psi_3\rangle = S|\psi_2\rangle$ with respect to the axis s is now $\pi - 2\phi$. The sequence of the two operations $-T$ (oracle) and S (Grover diffuser) has rotated the initial state by the angle $\pi - 2\phi$ towards the target state.

Number of Grover iterations and quantum supremacy

The geometric reasoning in the previous section has shown that a single application of an oracle and diffuser rotates the initial state by a certain angle towards the target state. The oracle/diffuser sequence must now be repeated until the target state is reached (or, since only an integer number of operations is possible until one is sufficiently close to the target state). This requirement determines the number of Grover iterations required.

The angle of rotation for an iteration is $\pi - 2\phi$; in total, the initial state must be rotated by the angle ϕ. Thus, for the number k of iterations required, we obtain the condition:

$$k\,(\pi - 2\phi) = \phi. \tag{6.28}$$

In order to calculate the concrete number of iterations for a given database size N, it is easier to go from ϕ to the complement angle $\theta = \frac{\pi}{2} - \phi$ (Fig. 6.10b). According to Eq. (6.25), we can write:

$$\langle s|t \rangle = \cos\left(\frac{\pi}{2} - \theta\right) = \sin\theta = \frac{1}{\sqrt{N}}. \tag{6.29}$$

For $\phi \approx \pi$, the angle θ is small so that $\sin\theta \approx \theta$, and accordingly $\theta \approx \frac{1}{\sqrt{N}}$. For large databases, where the searched state is only one of many components of the symmetric state, this approximation becomes very good. In the same approximation, let us assume that we can replace ϕ on the right-hand side of Eq. (6.28) by $\frac{\pi}{2}$ so that the total angle between the initial and target states is 90°. Then Eq. (6.28) becomes:

$$k \times 2\theta = \frac{\pi}{2}. \tag{6.30}$$

With $\theta \approx \frac{1}{\sqrt{N}}$, this gives $k \approx \sqrt{N} \times \frac{\pi}{4}$ iterations. Thus, the number k of iterations required for a given database size grows with the square root of the number of database entries. The Grover algorithm can, therefore, search a large unordered database faster than a classical search algorithm (which requires $\frac{N}{2}$ calls on average, so scales with N).

> *Quantum supremacy in database search*: As the database size increases, the number of iterations in the Grover algorithm scales with $\sqrt{N}$. This makes it more efficient for large databases than classical search algorithms, which scale with N.

With its unconventional and counterintuitive way of replacing a rotation with a sequence of reflections, the Grover algorithm does not seem elegant or particularly effective. Nevertheless, it can be shown to be optimal [78]; i. e., except for constant factors, there is no more efficient quantum algorithm.

6.4 Implementing the Grover algorithm with quantum gates

Let us now turn to the practical task of implementing the Grover algorithm on a quantum computer. To do this, we need to translate the unitary operations of the algorithm into actions of quantum gates. In classical programming, this is done by a compiler. However, since quantum computer programming is still largely done at the gate level, we will implement the algorithm by hand. This will allow us to gain useful experience with quantum gates.

We want to express the Grover diffuser and oracle by a sequence of standard quantum gates as implemented in available quantum computer systems (cf. p. 205). The problem is definitely solvable because, as unitary transformations, oracle and diffuser can always be represented by a sequence of universal quantum gates (e. g., CNOT).

Notation: When describing qubit operations on systems of multiple qubits, we need to specify which gate acts on which qubit. To do this, we use a notation in which the qubit (or qubits) in question is given as an index in parentheses: for example, $H_{(k)}$ means the application of the Hadamard gate to qubit k. The expression $H_{(1...k)}$ describes the application of H to qubits 1 to k, while $H_{(all)}$ represents the application of H to all qubits. In more theoretically oriented texts, this is often expressed by the notation $H^{\otimes n}$.

Implementation of the diffuser

First, we express the Grover diffuser $S = 2\,|s\rangle\,\langle s| - \mathbb{1}$ by quantum gates. We bring it into a more manageable form by expressing it in terms of standard quantum gates (mathematically: unitary transformations). First, we reduce the complicated superposition state $|s\rangle$ to the simpler state $|00\ldots0\rangle$. This is done by applying the Hadamard gate to all qubits. We take advantage of the fact that $H = H^\dagger = H^{-1}$ and consider the following transformed operator:

$$
\begin{aligned}
S' &= -H_{(all)} S H_{(all)} \\
&= -2\,\underbrace{H_{(all)}\,|s\rangle}_{=|00...0\rangle}\,\underbrace{\langle s|\,H_{(all)}}_{=\langle 00...0|} + \underbrace{H_{(all)}\,\mathbb{1}\,H_{(all)}}_{=\mathbb{1}} \\
&= \mathbb{1} - 2\,|00\ldots0\rangle\,\langle 00\ldots0|\,. \tag{6.31}
\end{aligned}
$$

In the second line, we used $|s\rangle = H_{(all)}\,|00\ldots0\rangle$, which generalizes Eq. (6.17), and $HH = \mathbb{1}$. We can read the effect of the operation S' thus obtained directly from Eq. (6.31): It leaves all components of the overall state unchanged (operator $\mathbb{1}$), except for $|00\ldots0\rangle$. For this component, the second term inverts the sign.

A look at the collection of standard quantum gates (p. 205) reveals two candidates that perform a similar operation (Fig. 6.12): The *CCZ gate* inverts the sign of the state component $|11\ldots1\rangle$, and the *Toffoli gate* swaps $|0\rangle$ and $|1\rangle$ at the last qubit for the state component that has $|1\rangle$ at all other qubits.

We see that the CCZ gate performs *almost* the desired operation, but with the state component $|11\ldots1\rangle$ instead of $|00\ldots0\rangle$. This problem can be solved by applying the X gate to all qubits. We have already considered this gate on p. 144; it performed the "penny flip" and swapped $|0\rangle$ and $|1\rangle$. So, we perform another transformation:

$$
S'' = X_{(all)} S' X_{(all)} = \mathbb{1} - 2\,|11\ldots1\rangle\,\langle 11\ldots1|\,. \tag{6.32}
$$

The equation is easily verified with $X = |0\rangle\,\langle 1| + |1\rangle\,\langle 0|$ and $X = X^\dagger = X^{-1}$. Thus, we have found an explicit realization of S'' by standard quantum gates. The right side of Eq. (6.32) is the representation of the CCZ gate in the Dirac notation:

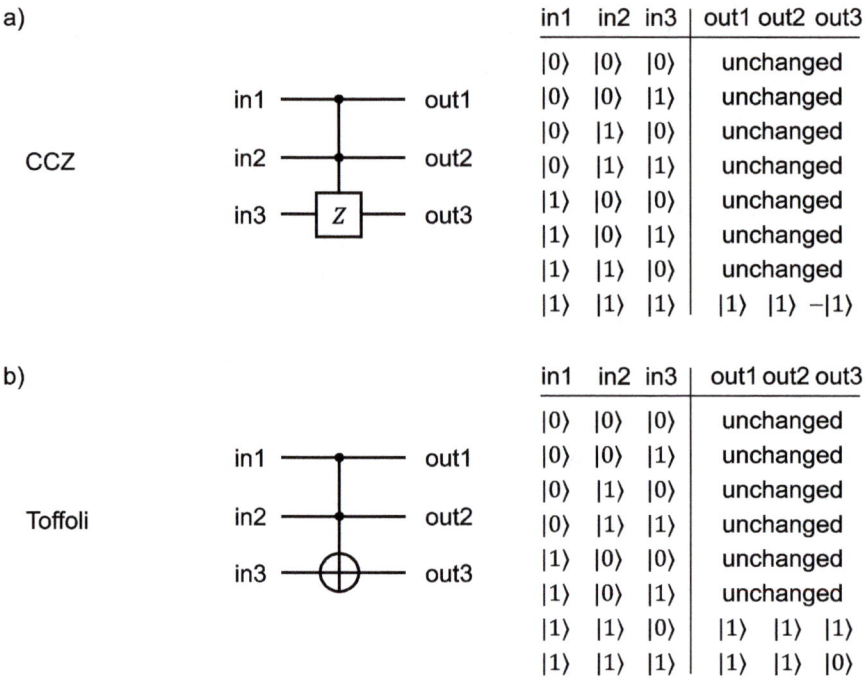

a)

in1	in2	in3	out1	out2	out3
$|0\rangle$	$|0\rangle$	$|0\rangle$	unchanged		
$|0\rangle$	$|0\rangle$	$|1\rangle$	unchanged		
$|0\rangle$	$|1\rangle$	$|0\rangle$	unchanged		
$|0\rangle$	$|1\rangle$	$|1\rangle$	unchanged		
$|1\rangle$	$|0\rangle$	$|0\rangle$	unchanged		
$|1\rangle$	$|0\rangle$	$|1\rangle$	unchanged		
$|1\rangle$	$|1\rangle$	$|0\rangle$	unchanged		
$|1\rangle$	$|1\rangle$	$|1\rangle$	$|1\rangle$	$|1\rangle$	$-|1\rangle$

b)

in1	in2	in3	out1	out2	out3
$|0\rangle$	$|0\rangle$	$|0\rangle$	unchanged		
$|0\rangle$	$|0\rangle$	$|1\rangle$	unchanged		
$|0\rangle$	$|1\rangle$	$|0\rangle$	unchanged		
$|0\rangle$	$|1\rangle$	$|1\rangle$	unchanged		
$|1\rangle$	$|0\rangle$	$|0\rangle$	unchanged		
$|1\rangle$	$|0\rangle$	$|1\rangle$	unchanged		
$|1\rangle$	$|1\rangle$	$|0\rangle$	$|1\rangle$	$|1\rangle$	$|1\rangle$
$|1\rangle$	$|1\rangle$	$|1\rangle$	$|1\rangle$	$|1\rangle$	$|0\rangle$

Figure 6.12: (a) The CCZ gate (Controlled-Controlled-Z) inverts the sign for the state component $|111\rangle$. (b) The Toffoli gate inverts the third qubit for the state component where both control bits are 1. Both gates are extendable to more than three qubits but then have to be assembled from simpler quantum gates in the development environments.

$$S'' = CCZ = \mathbb{1} - 2\,|11\ldots 1\rangle\,\langle 11\ldots 1|. \qquad (6.33)$$

To obtain the quantum gate realization of the Grover diffuser, we now perform the two transformations in reverse order. Starting from

$$S'' = X_{(all)}S'X_{(all)} = -X_{(all)}H_{(all)}SH_{(all)}X_{(all)}, \qquad (6.34)$$

we multiply from left and right by the inverse and use again $X = X^{-1}$ and $H = H^{-1}$:

$$S = -H_{(all)}X_{(all)}CCZX_{(all)}H_{(all)}. \qquad (6.35)$$

With this equation, we have achieved our objective of expressing the Grover diffuser $S = 2\,|s\rangle\,\langle s| - \mathbb{1}$ by standard quantum gates. All operators on the right-hand side of Eq. (6.35) are included in the quantum gate table on p. 205 and can be used in the available quantum computing programming environments. The global sign is irrelevant. Fig. 6.13 shows the graphical representation used in programming for the case of three qubits. The generalization to n qubits is obvious, although the CCZ gate for more than

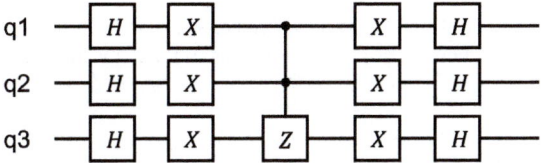

Figure 6.13: Realization of the Grover diffuser with standard quantum gates.

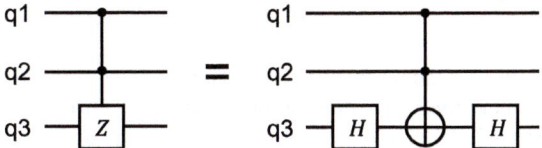

Figure 6.14: Realization of the CCZ gate with one Toffoli gate and two Hadamard gates.

three qubits is not part of the standard gate inventory and has to be emulated by a combination of other quantum gates.

Example: In some programming environments, the CCZ gate does not exist, but the Toffoli gate does. Show that the CCZ gate can be represented by the Toffoli gate using the arrangement shown in Fig. 6.14.

Solution: We consider only the case of three qubits; the generalization to n qubits is straightforward. To demonstrate the relation shown in Fig. 6.14, we consider the inverse relation (except for the sign):

$$\text{Toffoli} = -H_{(3)} CCZ H_{(3)}, \tag{6.36}$$

from which the desired relation can be obtained by multiplying $H_{(3)}$ from the left and right. With Eq. (6.11): $H_{(3)} = \frac{1}{\sqrt{2}}[|0\rangle \langle 0|_{(3)} + |1\rangle \langle 0|_{(3)} + |0\rangle \langle 1|_{(3)} - |1\rangle \langle 1|_{(3)}]$ and CCZ from Eq. (6.33), we simplify the right side, considering only those terms that are not zero from the outset:

$$\text{Toffoli} = \mathbb{1} - 2 \times \frac{1}{2}\left[|0\rangle \langle 1|_{(3)} - |1\rangle \langle 1|_{(3)}\right]|111\rangle \langle 111|\left[|1\rangle \langle 0|_{(3)} - |1\rangle \langle 1|_{(3)}\right]$$

$$= \mathbb{1} - \left[|110\rangle - |111\rangle\right]\left[\langle 110| - \langle 111|\right]$$

$$= \mathbb{1} - |110\rangle \langle 110| + |110\rangle \langle 111| + |111\rangle \langle 110| - |111\rangle \langle 111|. \tag{6.37}$$

The last equation looks complicated but is actually easy to read. The last four terms are only different from zero when applied to a state whose first two qubits are in state $|1\rangle$. Only for these state components the operator has an effect; for all others, it is equivalent to the unit operator $\mathbb{1}$. For the affected states, the last four terms add two entries to the unit matrix and subtract two entries. In matrix notation, for the third qubit only, the right-hand side is:

$$\mathbb{1} - \text{four terms} = \begin{pmatrix} 1 & 0 \\ 0 & 1 \end{pmatrix} + \begin{pmatrix} -1 & 1 \\ 1 & -1 \end{pmatrix} = \begin{pmatrix} 0 & 1 \\ 1 & 0 \end{pmatrix}. \tag{6.38}$$

This is indeed the matrix for the X gate (cf. Eq. (6.4)) acting on the third qubit when the first two qubits are in the state $|1\rangle$. This corresponds to the action of the Toffoli gate described in the truth table in Fig. 6.12, so we have shown the validity of the relation (6.36). The CCZ gate can thus be realized as shown in Fig. 6.14 using one Toffoli gate and two H gates. As a further illustration, we give the matrix representation for Toffoli and CCZ gates:

$$
\text{Toffoli:} \begin{pmatrix} 1 & 0 & \cdots & 0 & 0 \\ 0 & 1 & \cdots & 0 & 0 \\ \vdots & \vdots & \ddots & \vdots & \vdots \\ 0 & 0 & \cdots & 0 & 1 \\ 0 & 0 & \cdots & 1 & 0 \end{pmatrix}, \quad \text{CCZ:} \begin{pmatrix} 1 & 0 & \cdots & 0 & 0 \\ 0 & 1 & \cdots & 0 & 0 \\ \vdots & \vdots & \ddots & \vdots & \vdots \\ 0 & 0 & \cdots & 1 & 0 \\ 0 & 0 & \cdots & 0 & -1 \end{pmatrix}, \tag{6.39}
$$

For three qubits, these are 8×8 matrices because the state space has $2^3 = 8$ dimensions, and there are eight orthogonal basis states. The four entries in the lower right corner act on the state components $|110\rangle$ and $|111\rangle$, where the first two qubits are in the state $|1\rangle$ (cf. p. 143).

Function of the oracle and phase kickback

Having implemented the Grover diffuser using quantum gates, we now want to do the same for the oracle, the second important component in the Grover algorithm. While the diffuser is always the same, the oracle is problem-specific – it has to identify the particular states we are looking for. So, we can only consider special examples.

The oracle is supposed to invert the sign of the state component corresponding to the searched database entry. A technique often used in this context is called *phase kickback* and is so common in quantum computing that we will discuss it in more detail. First, we need the ancilla qubit shown in Fig. 6.9. Physically, it is not different from the other qubits. It is temporarily used as an "auxiliary register", but does not take part in the actual computation. In order not to destroy the unitary evolution of the other qubits, it has to be returned to its initial state at the end of the intermediate computation.

To keep the formulas simple, we return to our initial example with 2 qubits. For the present purpose, the ancilla qubit has to be prepared in the state $|1\rangle$ (which can be achieved by an X gate). This is the initial state of the whole system:

$$
|\psi_0\rangle = |0\rangle_1 |0\rangle_2 |1\rangle_a. \tag{6.40}
$$

Applying the Hadamard gate according to Eq. 6.9 brings all three qubits into a superposition state. Similar as in Eq. (6.19), the result after expanding is:

$$
|\psi_1\rangle = \frac{1}{2}\big[|00\rangle + |01\rangle + |10\rangle + |11\rangle\big] \times \frac{1}{\sqrt{2}}\big[|0\rangle_a - |1\rangle_a\big], \tag{6.41}
$$

where Eq. (6.7) was used. This is the state of the entire system before the oracle is applied.

When using phase kickback, the oracle works by "flipping" the state of the ancilla bit if the input is the searched state. Thus, it transforms $|0\rangle$ into $|1\rangle$ and vice versa. This operation can be realized by the Pauli-X gate. Thus, if $|01\rangle$ is the searched state, then the total state of the system after applying the oracle is:

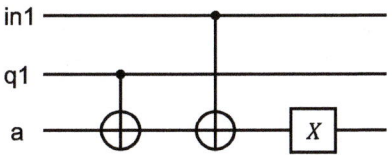

Figure 6.15: Comparing two qubits.

$$|\psi_2\rangle = \frac{1}{2}\big[|00\rangle + |10\rangle + |11\rangle\big] \times \underbrace{\frac{1}{\sqrt{2}}\big[|0\rangle_a - |1\rangle_a\big]}_{\text{unchanged} = \text{no}} + \frac{1}{2}\,|01\rangle \times \underbrace{\frac{1}{\sqrt{2}}\big[|1\rangle_a - |0\rangle_a\big]}_{\text{flipped} = \text{yes}}\,.$$

By extracting the sign from the last parenthesis, it can be written as follows:

$$|\psi_1\rangle = \frac{1}{2}\big[|00\rangle - |01\rangle + |10\rangle + |11\rangle\big] \times \frac{1}{\sqrt{2}}\big[|0\rangle_a - |1\rangle_a\big]. \tag{6.42}$$

By applying the Hadamard gate again, the ancilla qubit is restored to its initial state $|1\rangle$. It no longer participates in the following computation. The state of the entire system is now:

$$|\psi_1\rangle = \frac{1}{2}\big[|00\rangle - |01\rangle + |10\rangle + |11\rangle\big]\,|1\rangle_a. \tag{6.43}$$

If we now ignore the ancilla qubit again (which is allowed because in Eq. (6.43), it is not entangled with any of the other qubits and is independent of their states), we have generated the state for the qubits q_1 and q_2 that we assumed in Eq. (6.20).

Due to the action of the oracle on the ancilla qubit, the sign of the searched state $|01\rangle$ has changed in the overall state, although the oracle does not interact with the qubits q_1 and q_2 at all. Such an indirect sign or phase change by interaction with an ancilla qubit is called *phase kickback* and is widely used in quantum algorithms.

Example – comparing two qubits: As a preliminary consideration for the next section, show that the circuit in Fig. 6.15 can be used to compare the two qubits in_1 and q_1. The ancilla qubit is initially prepared in the state $|0\rangle$. At the end, it should be in the state $|1\rangle$ for those state components for which $in_1 = q_1$ and in the state $|0\rangle$ for those components with $in_1 \neq q_1$.

Solution: We write the states of the three qubits in the order $|in_1, q_1, a\rangle$. The ancilla qubit is prepared in the state $|0\rangle$ at the beginning; for the state of the other two qubits, we generally assume:

$$|in_1\rangle = \alpha\,|0\rangle + \beta\,|1\rangle \quad \text{and} \quad q_1 = \gamma\,|0\rangle + \delta\,|1\rangle\,. \tag{6.44}$$

By choosing $\alpha = 1$ or $\beta = 1$, the input qubit can be set to the state $|0\rangle$ or $|1\rangle$. Thus, the initial state of the system is:

$$|\psi_0\rangle = \alpha\gamma\,|000\rangle + \alpha\delta\,|010\rangle + \beta\gamma\,|100\rangle + \beta\delta\,|110\rangle\,. \tag{6.45}$$

The goal is to set the last qubit to $|1\rangle$ if the first two qubits are equal, i.e., for the state components $|000\rangle$ and $|110\rangle$ by applying the sequence of quantum gates shown in Fig. 6.15. We first apply the CNOT gate to q_1

and a. The corresponding truth table is shown in Fig. 6.3. The CNOT gate flips the third qubit for the state components where the second qubit is equal to $|1\rangle$ and has no effect on the other state components. This results in the state $|\psi_1\rangle = \text{CNOT}_{(q_1,a)} |\psi_0\rangle$:

$$|\psi_1\rangle = \alpha\gamma\,|000\rangle + \alpha\delta\,|0\underline{11}\rangle + \beta\gamma\,|100\rangle + \beta\delta\,|1\underline{11}\rangle\,. \tag{6.46}$$

We have underlined the qubits where "something happens". Now we apply CNOT to in_1 and a. The third qubit is flipped for the state components where the first qubit is equal to $|1\rangle$:

$$|\psi_2\rangle = \text{CNOT}_{(\text{in}_1,a)} |\psi_1\rangle = \alpha\gamma\,|000\rangle + \alpha\delta\,|011\rangle + \beta\gamma\,|\underline{10}1\rangle + \beta\delta\,|\underline{110}\rangle\,. \tag{6.47}$$

The ancilla bit is now equal to $|1\rangle$ if $\text{in}_1 \neq q_1$ and $|0\rangle$ otherwise. This is exactly the opposite of what we wanted. We correct this by flipping the ancilla qubit with the X gate:

$$|\psi_3\rangle = X_{(3)} |\psi_2\rangle = \alpha\gamma\,|001\rangle + \alpha\delta\,|010\rangle + \beta\gamma\,|100\rangle + \beta\delta\,|111\rangle\,. \tag{6.48}$$

This realizes the desired state: The ancilla qubit indicates with a $|1\rangle$ that in_1 and q_1 are equal; it is $|0\rangle$ if they are not equal.

Implementation of an oracle: *"His Master's Voice"*

At this point, we have to deal with a circumstance that often leads to irritation. When discussing the Grover algorithm, it is often assumed that the algorithm needs to know the searched state $|t\rangle$ in advance in order to work. The crucial task of identifying the searched state is delegated to the oracle, but the algorithm makes no statement about how the oracle works. In our discussion of phase kickback, we also had to identify the searched state "by hand". We want to remove this difficulty of understanding by implementing a simple oracle in detail at the qubit level.

Our oracle identifies the searched state based on user input. There are two input qubits, in_1 and in_2, which the user can arbitrarily set to $|0\rangle$ or $|1\rangle$ with an X gate. The oracle marks the state indicated in this way. If the user prepares $\text{in}_1 = |0\rangle$ and $\text{in}_2 = |0\rangle$, then $|00\rangle$ is regarded as the searched state. For $\text{in}_1 = |0\rangle$ and $\text{in}_2 = |1\rangle$, it is $|01\rangle$. Because the oracle does nothing but faithfully follow the user's instructions, we call it "His Master's Voice" (HMV). To fully implement Grover's algorithm with the HMV oracle, we just need to collect what we have achieved so far. Fig. 6.16 shows the process. We need 7 qubits:

- two input qubits in_1 and in_2 for user input,
- two qubits q_1 and q_2 numbering the states and performing the actual Grover algorithm,
- two ancilla qubits a_1 and a_2 for the qubit comparison according to Fig. 6.15,
- another ancilla qubit a_3 for the phase kickback.

Lines 1–3 and 4–6 in Fig. 6.16 correspond to each other and would be repeated more often in a larger database. The last line contains the ancilla qubit a_3, which is used to perform the phase kickback at the end. User input is provided by the two input qubits in_1 and in_2.

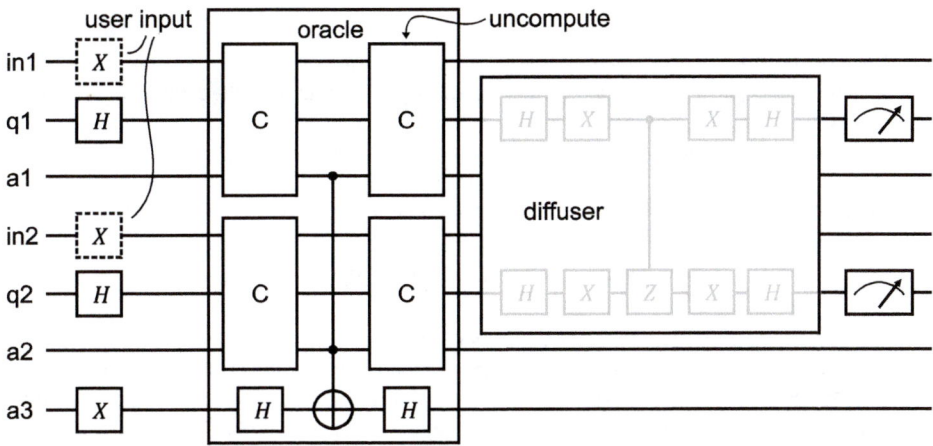

Figure 6.16: Implementation of the Grover algorithm using the HMV oracle.

They can be switched by the user from the initial state $|0\rangle$ to the state $|1\rangle$ on demand by the dashed X gates. The remaining quantum gates in the left column are used to initialize the corresponding qubits as described above.

The oracle labels the state components specified by the user input. This is done by performing the qubit comparison described above for both input qubits. The boxes labeled C (*Compare*) correspond to the gate combination from Fig. 6.15. Two qubits are compared at a time (in$_1$ with q$_1$ and in$_2$ with q$_2$). The result is stored in the ancilla qubits a$_1$ and a$_2$.

The searched state component is the one with in$_1$ = q$_1$ and in$_2$ = q$_2$. For this component, both ancilla qubits a$_1$ and a$_2$ are in the state $|1\rangle$ after the C blocks. This is checked with the Toffoli gate. It flips the ancilla qubit a$_3$ if the condition is fulfilled and thus performs a phase kickback. The searched state component is now marked with a minus sign, as desired. To complete the oracle, the ancilla qubits a$_1$ and a$_2$ must be returned to their initial states. This is done in the column labeled "uncompute" by executing the two C blocks again.

This completes the oracle, and the diffuser can take over. Its operation has already been discussed. For a database with two qubits, a single pass is sufficient to complete the algorithm. For larger databases, the sequence of oracle and diffuser would have to be repeated several times.

Example – the HMV oracle in detail: Convince yourself that the HMV oracle really works as described by tracing the state of the qubits step by step through the algorithm. Use $|q_1, q_2\rangle = |01\rangle$ as the searched state. **_i_**

Solution: Let us go through the algorithm step by step using Fig. 6.16. We specify the state of the qubits in the order $|in_1, q_1, a_1, in_2, q_2, a_2, a_3\rangle$ (i. e., from top to bottom). At the beginning of the computation, all qubits are in the state $|0\rangle$.

1. *Step 1: Initialization*
   The X and H gates in the left column are used to initialize the qubits:
   - The user input is assumed to be $in_1 = |0\rangle$ and $in_2 = |1\rangle$. Thus, only the second of the dashed X gates is needed.
   - The Hadamard gates put q_1 and q_2 into the symmetric superposition state (6.19).
   - The ancilla qubit a_3 is put into the state $|1\rangle$ with an X gate, which is needed for the phase kickback.
   The overall state of the system after this initialization phase is:

   $$|\psi_1\rangle = \frac{1}{2}\Big[|0001001\rangle + |0001101\rangle + |0_{in_1}1_{q_1}01_{in_2}0_{q_2}01\rangle + |0101101\rangle\Big]. \tag{6.49}$$

   This state has four terms corresponding to the symmetric superposition of q_1 and q_2. At the fourth position (qubit in_2), there is a 1 in all terms because this qubit was turned on with the corresponding X gate. For clarity, the names of the important qubits are explicitly appended in the third term. The second term is the one where in_1/q_1 and in_2/q_2 match. It has to be found by the algorithm.

2. *Step 2: Hadamard gate on ancilla qubit a_3*
   To perform the phase kickback, the ancilla qubit a_3 must be set to state

   $$|-\rangle = \frac{1}{\sqrt{2}}\Big[|0\rangle - |1\rangle\Big] \tag{6.50}$$

   (cf. Eq. (6.41)). This is done by the Hadamard gate in the last line of Fig. 6.16. We do not express this state in the computational basis ($|0\rangle$, $|1\rangle$) but insert it directly in the above notation:

   $$|\psi_2\rangle = \frac{1}{2}\Big[|000100-\rangle + |000110-\rangle + |010100-\rangle + |010110-\rangle\Big]. \tag{6.51}$$

3. *Step 3: Perform the comparisons with the input qubits*
   Next, the equality test described in the previous example is performed for both qubit pairs in_1/q_1 and in_2/q_2. It checks for all state components whether $in_1 = q_1$ and $in_2 = q_2$ and stores the result in a_1 and a_2. The comparison is made with the gate combination C from Fig. 6.15, which flips the corresponding ancilla qubit if the two qubits being compared are in the same state. Applying C to the first qubit pair in_1 and q_1 yields the overall state:

   $$|\psi_3\rangle = \frac{1}{2}\Big[|\underline{00}1100-\rangle + |\underline{00}1110-\rangle + |\underline{01}0100-\rangle + |\underline{01}0110-\rangle\Big]. \tag{6.52}$$

   The two qubits being compared are underlined; a flipped qubit is marked with an arrow. Applying C to the second qubit pair in_2 and q_2 results in

   $$|\psi_4\rangle = \frac{1}{2}\Big[|001\underline{10}0-\rangle + |001_{a_1}\underline{11}1_{a_2}-\rangle + |010\underline{10}0-\rangle + |010\underline{11}1-\rangle\Big]. \tag{6.53}$$

   Note that for the second component of the total state, both ancilla qubits a_1 and a_2 (labeled for clarity) have the value 1 after the two C blocks. This is exactly the component we want to find.

4. *Step 4: Toffoli gate and phase kickback*
   The Toffoli gate acts as an X gate for a_3 in the component where both ancilla qubits a_1 and a_2 have the value 1, i. e., the second. All other components are unchanged. With $X|-\rangle = -|-\rangle$, we get the negative sign in the second term that is needed for the algorithm:

   $$|\psi_5\rangle = \frac{1}{2}\Big[|001100-\rangle - |001111-\rangle + |010100-\rangle + |010111-\rangle\Big]. \tag{6.54}$$

5. *Step 5: Uncompute*

   The ancilla qubits a_1 and a_2 are not needed for the further computation. However, they are entangled with the other qubits in the state (6.54). If we would simply ignore them in the following and continue computing with the remaining qubits alone, we would have entanglement for this reduced system that extends beyond the boundaries of the system considered – a feature of decoherence that prevents interference from occurring. For the reduced system to evolve unitary and be capable of interference, the information contained in the ancilla qubits must be "erased" and the entanglement removed.

   This process is called *uncomputation* (cf. p. 141). Since one cannot simply "delete" a_1 and a_2 in quantum computing (that would be a nonunitary operation), one simply runs through the computational steps made with these bits backwards, thus restoring them to their initial state. This is done with the two C blocks to the right of the Toffoli gate, as $C^{-1} = C$. After the upper C block, the state is:

   $$|\psi_6\rangle = \frac{1}{2}\Big[|\underset{\uparrow}{000}100-\rangle - |\underset{\uparrow}{000}111-\rangle + |\underline{010}100-\rangle + |\underline{010}111-\rangle\Big]. \tag{6.55}$$

   The second C block is passed, and with a H gate on a_3, we return this qubit to the initial state $|1\rangle$:

   $$|\psi_7\rangle = \frac{1}{2}\Big[|0001\underline{001}\rangle - |0001\underset{\uparrow}{1}01\rangle + |0101\underline{001}\rangle + |0101\underset{\uparrow}{1}01\rangle\Big]. \tag{6.56}$$

   Now all ancilla bits are in well-defined states and not entangled with other qubits. For the input qubits, this was the case from the beginning. Therefore, we can write the state $|\psi_7\rangle$ as a product state as follows:

   $$|\psi_7\rangle = \frac{1}{2}\Big[|00\rangle - |01\rangle + |10\rangle + |11\rangle\Big]_{q_1,q_2} \times |0_{in_1} 1_{in_2} 0_{a_1} 0_{a_2} 1_{a_3}\rangle. \tag{6.57}$$

   All together we have created a state as in Eq. (6.43), where the qubits q_1 and q_2 are in a superposition state, and the component we are looking for is marked by its sign. It will be selected by the Grover diffuser in the next step. The remaining qubits are no longer entangled with q_1 and q_2 and therefore do not affect the subsequent calculation.

Uncomputation and decoherence caused by the environment

The previous calculation gives us reason to consider again the concept of environmental decoherence and the associated need for uncomputing. We have already discussed in Section 3.10 how entanglement extending beyond the boundaries of the system under consideration can prevent interference in the system. Let us illustrate this again with a concrete example. The system we consider (and on which we later make measurements) are the qubits $|q_1\rangle$ and $|q_2\rangle$. The "environment", which we are not interested in, consists of the ancilla qubits $|a_{1,2,3}\rangle$ and the input qubits $|in_{1,2}\rangle$. The division into system and environment results from which constituents we actively track (they form the system) and which we ignore in subsequent measurements (they form the environment).

The state (6.54) cannot be expressed as the product of $|system\rangle \times |environment\rangle$, i. e., in the form $|q_1, q_2\rangle \times |a_1, a_2, a_3, in_1, in_2\rangle$. This means that the system and its environment are entangled. We have already discussed on p. 84 what entanglement across system boundaries means for measurements on system variables: If "which-way information" (or more generally: "which-state information") leaks to the environment, no interference occurs in measurements of the corresponding system variables.

Since interference is essential to the operation of quantum algorithms, entanglement across the boundaries of the qubit system must either be prevented or reversed. One could formulate a quantum version of the "Vegas Rule": "*What happens in Vegas has to stay in Vegas*". The information must not leak out; otherwise, the ability to interfere will be destroyed.

Environment-induced decoherence in quantum computing: There must be no feature in the environment from which information about the state of the qubits can be read. If this happens, interference will be suppressed, and any quantum advantage will be lost.

6.5 Quantum Fourier Transform

The *Quantum Fourier Transform (QFT)* is one of the most important tools in quantum computing. It was discovered in connection with the development of Shor's algorithm by Dan Coppersmith in 1994 [79]. The QFT is used to reveal periodicities in bit sequences and does so – compared to its classical counterpart – with a significant advantage in computational cost. Thus, the quantum Fourier transform is the basis for many quantum algorithms. The most prominent example is Shor's algorithm for factoring large numbers. The core idea of this algorithm, which we will discuss in more detail in the next section, is to reduce factorization to the detection of periodicities and then apply the Quantum Fourier Transform.

Classical Fourier transform

The *Fourier transform* is one of the most important algorithms in classical physics and computer science. To get a feel for its power, we will first examine the classical algorithm and how it works. The transfer to the quantum domain is then relatively straightforward.

The Fourier transform expresses a function $x(t)$ as a sum or integral of periodic functions with different frequencies. It decomposes the function into its frequency components and thus represents it in terms of sine, cosine, or complex exponential functions. The latter is the easiest to handle in calculations and is therefore often used. A function $x(t)$ can be written as a superposition of complex exponential functions $e^{-2\pi i f t}$ with frequency f as follows:

$$x(t) = \frac{1}{\sqrt{2\pi}} \int_{-\infty}^{+\infty} df \, y(f) \, e^{-2\pi i f t}. \tag{6.58}$$

The *Fourier coefficients* $y(f)$ indicate the weight of the frequency f in the integral. They are usually complex and can be calculated using the following equation:

$$y(f) = \frac{1}{\sqrt{2\pi}} \int_{-\infty}^{+\infty} dt \, x(t) \, e^{2\pi i f t}. \tag{6.59}$$

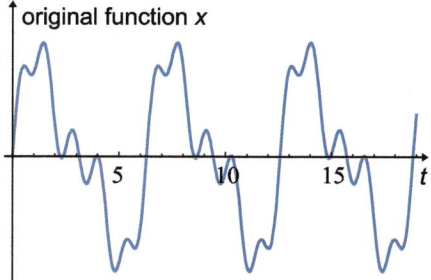
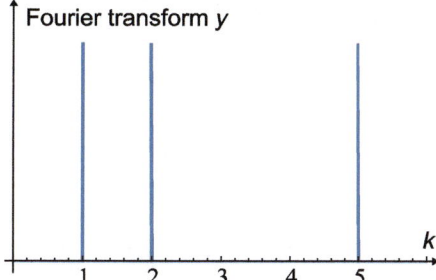

Figure 6.17: Fourier transform of a periodic function.

The symmetrically looking pair of equations (6.58) and (6.59) describe the Fourier transform and its inverse. There are several conventions for defining the Fourier transform. Here, the equations have been written to make the transition to the discrete and to the Quantum Fourier Transform as easy as possible. In mathematics and physics, the sign of the exponent is often reversed.

Fig. 6.17 shows a periodic function $x(t)$ constructed as the sum of three sine functions. It is decomposed into its frequency components by the Fourier transform. The Fourier transform $y(k)$ shows sharp maxima at the three corresponding frequencies.

    If you record the sound of a musical instrument with a microphone and look at the resulting signal, you will get curves very similar to the function $x(t)$ shown. Musical instruments have a basic tone of a certain frequency (e. g., 440 Hz = concert pitch a) and additional harmonics that are multiples of that basic frequency – similar to the spectrum in Fig. 6.17 on the right.

    The Fourier transform and its inverse, provide a simple way to switch the mathematical description of a signal between the time and frequency domains. This is used in many applications, such as audio or video signal processing.

Discrete Fourier transform

In the discrete Fourier transform (DFT), the same concept is applied to discrete sequences of numbers. This is necessary, for example, when the function is stored on a computer, and the Fourier transform is to be performed numerically. Instead of a continuous function $x(t)$, a sequence of numbers $(x_0, \ldots, x_{N-1})$ is considered. This vector is mapped to the Fourier transform $(y_0, \ldots, y_{N-1})$, which is also discrete. The formula for calculating the discrete Fourier transform is the direct analog of Eq. (6.59):

$$y_k = \frac{1}{\sqrt{N}} \sum_{j=0}^{N-1} x_j \, e^{2\pi i \frac{jk}{N}} . \tag{6.60}$$

Comparing the two formulas, we see that the frequency is $f = k/N$ and the period is $T = 1/f = N/k$.

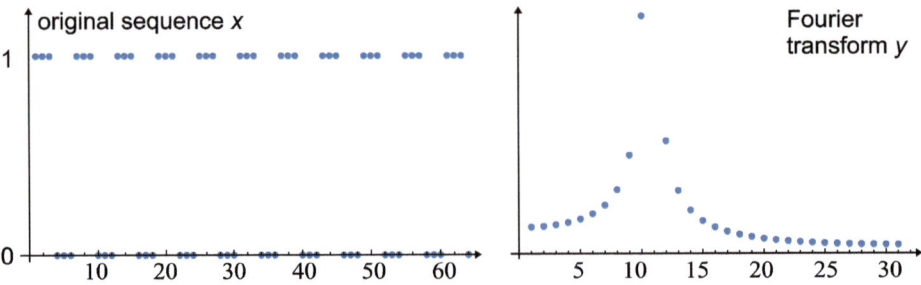

Figure 6.18: Discrete Fourier transform of a periodic bit sequence.

ℹ️ *Leakage in the discrete Fourier transform:* Although the discrete Fourier transform is defined for any number format that can be represented on a computer, we consider a periodic bit sequence of zeros and ones in Fig. 6.18. It has a length of $N = 64$, and its period is 6. The modulus of the Fourier coefficients is shown on the right.

The maximum at $k = 10$ reflects the periodicity of the bit sequence (since $T = N/k$ leads to $k = N/T = 64/6 \approx 10$). Despite the periodicity of the initial sequence, there are nonvanishing Fourier components not only at $k = 10$, but also at neighboring values. This is a peculiarity of the DFT: since N is not an integer multiple of the period, the algorithm experiences apparent "jumps": deviations from the periodicity at the edge of the number sequence. Mathematically, the effect is a "spillover" of the amplitudes to neighboring frequencies. This effect is called *leakage*. We will also encounter it in Shor's algorithm. It has nothing to do with quantum physics but is a characteristic of the discrete Fourier transform.

ℹ️ *Example:* We consider a periodic sequence of numbers of length N with period r. To avoid the leakage effect, let N be a multiple of r. Show that nonzero Fourier coefficients y_k can only occur for those k which are integer multiples of N/r.

Solution: In the formula for calculating the Fourier components,

$$y_k = \frac{1}{\sqrt{N}} \sum_{j=0}^{N-1} x_j \, e^{2\pi i \frac{kj}{N}}, \tag{6.61}$$

we exploit the periodicity of the coefficients x_j. After every r terms, the coefficients x_j repeat. Therefore, we can decompose the total sum into partial sums. The first partial sum contains all terms proportional to $x_0 = x_r = x_{2r} = \cdots$, the second contains all terms proportional to $x_1 = x_{r+1} = x_{2r+1} = \cdots$:

$$y_k = \frac{1}{\sqrt{N}} \left[x_0 \sum_{j=0}^{\frac{N}{r}-1} e^{2\pi i \frac{krj}{N}} + x_1 \sum_{j=0}^{\frac{N}{r}-1} e^{2\pi i \frac{k\times(rj+1)}{N}} + \cdots + x_{r-1} \sum_{j=0}^{\frac{N}{r}-1} e^{2\pi i \frac{k\times(rj+r-1)}{N}} \right].$$

Each of the partial sums has only N/r terms, and the individual partial sums no longer depend on the values of x_j. Therefore, we can compute them explicitly. First, we extract from the partial sums all factors that do not depend on j and add them up:

$$y_k = \frac{1}{\sqrt{N}} \left[x_0 + x_1 e^{2\pi i \frac{k\times1}{N}} + x_2 e^{2\pi i \frac{k\times2}{N}} + \cdots + x_{r-1} e^{2\pi i \frac{k(r-1)}{N}} \right] \sum_{j=0}^{\frac{N}{r}-1} e^{2\pi i \frac{k\times rj}{N}}.$$

We write the terms in square brackets in mathematical sum notation, while for the sum on the right, we exploit the following important mathematical identity known as *orthogonality relation for the complex exponential function*:

Orthogonality relation for the complex exponential function:

$$\sum_{j=0}^{M-1} e^{\frac{2\pi i}{M} jk} = \begin{cases} M & \text{if } k \text{ is an integer multiple of } M, \\ 0 & \text{else.} \end{cases} \tag{6.62}$$

In our case, $M = N/r$, leading to:

$$y_k = \frac{1}{\sqrt{N}} \times \frac{N}{r} \times \sum_{n=0}^{r-1} x_n e^{2\pi i \frac{kn}{N}}, \quad \text{if } k \text{ is an integer multiple of } N/r, \tag{6.63}$$

and $y_k = 0$ otherwise. This proves the assertion made above. The sum term resulting from the square brackets is nothing else than a "small" DFT, covering only one period of the initial sequence.

Unitarity of the discrete Fourier transform

For purposes of numerical computation, it is helpful to interpret the discrete Fourier transform as a matrix operation that maps a vector x_j to a vector y_k:

$$y_k = \sum_{j=0}^{N-1} W_{kj} x_j, \tag{6.64}$$

with the matrix

$$W_{kj} = \frac{1}{\sqrt{N}} \omega^{jk}, \quad \text{where} \quad \omega^{jk} = e^{2\pi i \frac{jk}{N}}. \tag{6.65}$$

In matrix notation, this representation is transparent and symmetrical:

$$W = \frac{1}{\sqrt{N}} \begin{pmatrix} 1 & 1 & 1 & 1 & \cdots & 1 \\ 1 & \omega & \omega^2 & \omega^3 & \cdots & \omega^{N-1} \\ 1 & \omega^2 & \omega^4 & \omega^6 & \cdots & \omega^{2(N-1)} \\ 1 & \omega^3 & \omega^6 & \omega^9 & \cdots & \omega^{3(N-1)} \\ \vdots & \vdots & \vdots & \vdots & \ddots & \vdots \\ 1 & \omega^{N-1} & \omega^{2(N-1)} & \omega^{3(N-1)} & \cdots & \omega^{(N-1)(N-1)} \end{pmatrix}. \tag{6.66}$$

The matrix W is unitary because $W W^\dagger = \mathbb{1}$. The adjoint (i. e., the transposed and complex conjugated) matrix is equal to the inverse. To show this, we write:

$$W W^\dagger = \sum_j W_{nj} W_{mj}^* = \frac{1}{N} \sum_j e^{2\pi i \frac{nj}{N}} e^{-2\pi i \frac{mj}{N}} = \frac{1}{N} \sum_{j=0}^{N-1} e^{\frac{2\pi i}{N} j(n-m)}.$$

It follows directly from the orthogonality relation (6.62) with $k = n - m$ that the non-diagonal elements $(n \neq m)$ of $W\,W^\dagger$ are zero, while the diagonal elements $(n = m)$ are equal to 1. Thus, $W\,W^\dagger = \mathbb{1}$, and W is unitary.

This result is important for the quantum version of the Fourier transform, which is represented in the computational basis by the same matrix as Eq. (6.66). The fact that $W\,W^\dagger = \mathbb{1}$ implies that the quantum Fourier transform is a unitary operation. It is, therefore, reversible and can be represented by universal quantum gates.

Fast Fourier Transform (FFT)

What is the computational complexity of the discrete Fourier transform? Looking at Eq. (6.64), one would expect on the order of $\mathcal{O}(N^2)$ elementary operations, that is, scaling by N^2. This is because to compute y; the matrix W has to be multiplied by x. Since W has $N \times N$ complex-valued entries, at least N^2 complex multiplications are needed.

However, there is a much more efficient way to compute the DFT already in classical computer science. The *Fast Fourier Transform (FFT)*, known since the mid-1960s, is one of the most important numerical tools and is used for a wide range of purposes. It can be performed with the order of $\mathcal{O}(N \log_2 N)$ elementary arithmetic operations. This is a huge improvement: for example, if $N = 10^6$, the execution time is reduced by a factor of 50 000. This corresponds to a reduction from 13.9 hours to 1 second.

In the FFT, the number sequence is cut in half recursively. Therefore, bit sequences of length $N = 2^n$ are favorable for its implementation. The FFT is based on the symmetry of the matrix W. In Eq. (6.66) there are only a limited number of different entries, whose values are repeated several times because:

$$\omega^N = \left(e^{\frac{2\pi i}{N}}\right)^N = 1 \quad \text{and} \quad \omega^{\frac{N}{2}} = \left(e^{\frac{2\pi i}{N}}\right)^{\frac{N}{2}} = -1. \tag{6.67}$$

This symmetry is ingeniously exploited in the FFT. First, it can be shown that a discrete Fourier transform of length N can be reduced to the sum of two discrete Fourier transforms of length $N/2$ (lemma of Danielson and Lanczos). This can be done recursively so that after n steps only a matrix of size 1 remains. This reduces the Fourier transform to a simple multiplication of complex numbers, which afterwards have to be summed according to an elaborate scheme. The latter is done by cleverly swapping the bit order.

Transfer to the quantum domain: Quantum Fourier Transform

The *Quantum Fourier Transform (QFT)* is a straight translation of the discrete Fourier transform into the quantum domain.

Quantum Fourier Transform: The quantum Fourier transform transforms an initial state $|x\rangle$ into a final state $|y\rangle$:

$$|x\rangle = \sum_{j=0}^{N-1} x_j\,|j\rangle \longrightarrow |y\rangle = \sum_{k=0}^{N-1} y_k\,|k\rangle, \tag{6.68}$$

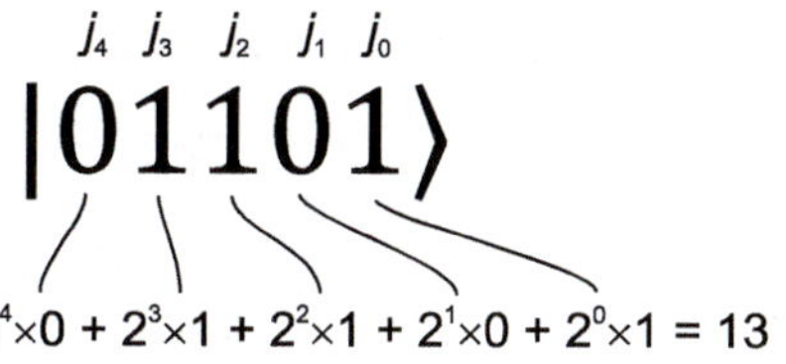

$$\overset{j_4\ j_3\ j_2\ j_1\ j_0}{|01101\rangle}$$

$$j = 2^4 \times 0 + 2^3 \times 1 + 2^2 \times 1 + 2^1 \times 0 + 2^0 \times 1 = 13$$

Figure 6.19: Binary representation and bit order in the quantum Fourier transform. Qubit 4 is the most significant one, Qubit 0 is the least significant.

where:

$$y_k = \frac{1}{\sqrt{N}} \sum_{j=0}^{N-1} x_j\, e^{2\pi i \frac{j\cdot k}{N}}. \qquad (6.69)$$

The coefficients x_j and y_k are classical numbers. The relationship between them is the same in the discrete Fourier transform in Eq. (6.60).

The equations (6.60) and (6.69) for the discrete and quantum Fourier transform are identical. Only Eq. (6.68) shows that there is a completely different concept behind the QFT. The notation must be explained to understand the underlying differences in the classical case.

The classical discrete Fourier transform acts on a vector with $N = 2^n$ entries, while the quantum Fourier transform acts on a register of n qubits. This quantum system can be described by $N = 2^n$ orthogonal basis states, denoted by $|j\rangle$ in Eq. (6.68). The coefficients x_j are the weights with which these basis states contribute to the total state. The QFT acts on all basis states simultaneously. We can write symbolically:

$$\mathbf{QFT}_N \begin{pmatrix} x_0\,|00\ldots000\rangle \\ +x_1\,|00\ldots001\rangle \\ +x_2\,|00\ldots010\rangle \\ +x_3\,|00\ldots011\rangle \\ \vdots \\ +x_{N-1}\,|11\ldots111\rangle \end{pmatrix} = \begin{pmatrix} y_0\,|00\ldots000\rangle \\ +y_1\,|00\ldots001\rangle \\ +y_2\,|00\ldots010\rangle \\ +y_3\,|00\ldots011\rangle \\ \vdots \\ +y_{N-1}\,|11\ldots111\rangle \end{pmatrix}. \qquad (6.70)$$

The Fourier coefficients y_k are the weights of the corresponding basis states in the final state. Both x_j and y_k are classical numbers. The relation (6.69) between them is the same as in the DFT.

Another peculiarity of the notation: In Eq. (6.68), the basis states are numbered from 0 to $N - 1$. For a unique assignment to the N basis states, one specifies that $|j\rangle$ denotes the basis state whose bit sequence is the binary representation of the number j. For example, the state $|01101\rangle$ is also written as $|13\rangle$ because 1101 is the binary representation of the number 13 (Fig. 6.19). As with any notation, there is no physics behind it. It has just proven to be convenient.

The symbolic representation in Eq. (6.70) shows the crucial advantage of QFT that makes it one of the mainstays of quantum computing: all components $|00\ldots000\rangle$ to $|11\ldots111\rangle$ are processed simultaneously by a single call of the QFT – a model example of quantum parallelism.

The disadvantage of quantum parallelism is equally obvious: with a single measurement in the computational basis, it is not possible to read out all coefficients y_k at once; only one particular final state is found. Because of the statistical nature of the measurement, the Fourier coefficients can only be deduced from the frequency distribution of the measurement results after many runs – but then the quantum advantage is gone.

Therefore, the quantum Fourier transform can only show its advantage for problems where not all components of the spectrum are individually relevant but where global properties are important. An example is the search for periodicities in number sequences, which is done in Shor's algorithm using the QFT.

Example: Give a representation of the QFT in the Dirac notation.

Solution: We take advantage of the fact that the relations (6.69) and (6.60) have exactly the same form for the coefficients of the DFT and the QFT. For the DFT, we have already found a matrix representation with Eq. (6.66). We can transfer it directly using Dirac's notation with $|k\rangle \langle j|$ denoting the entry in row k and column j:

$$\text{QFT}_N = \frac{1}{\sqrt{N}} \sum_{j=0}^{N-1} \sum_{k=0}^{N-1} \omega^{jk} |k\rangle \langle j|. \tag{6.71}$$

Product representation of the QFT

To run the QFT on a quantum computer, we have to express it in terms of quantum gates. The unitarity of QFT ensures that this is possible in principle. We have demonstrated unitarity above for the DFT; it transfers directly to the QFT.

The total state of the system consists of a register of n qubits. The QFT transforms the initial state $|x\rangle$ into the final state $|y\rangle$ according to Eq. (6.68). To realize this process by quantum gates, we need to determine how the QFT acts on each qubit individually. This is facilitated by the linearity of the QFT: it is sufficient to know its effect on the states of the computational basis, i. e. to consider the cases where only one of the x_j in Eq. (6.68) is different from zero. The general case is then followed by superposition.

Example: Consider a system of two qubits and explain how the QFT acts on the states of the computational basis.

Solution: We again use the notation $\omega = e^{\frac{2\pi i}{N}}$ and write Eq. (6.69) as follows:

$$y_k = \frac{1}{\sqrt{N}} \sum_{j=0}^{N-1} x_j \, \omega^{jk}. \tag{6.72}$$

For two qubits, the states of the computational basis are:

$$|0\rangle = |00\rangle, \quad |1\rangle = |01\rangle, \quad |2\rangle = |10\rangle, \quad |3\rangle = |11\rangle. \tag{6.73}$$

We evaluate Eq. (6.72) for each of them in turn. In each case, only one of the x_j is different from zero:

$$|00\rangle \xrightarrow{\text{QFT}} \frac{1}{2}\Big[|00\rangle + |01\rangle + |10\rangle + |11\rangle\Big] \qquad (x_0 = 1, \text{all others} = 0),$$

$$|01\rangle \xrightarrow{\text{QFT}} \frac{1}{2}\Big[|00\rangle + \omega\,|01\rangle + \omega^2\,|10\rangle + \omega^3\,|11\rangle\Big] \qquad (x_1 = 1, \text{all others} = 0),$$

$$|10\rangle \xrightarrow{\text{QFT}} \frac{1}{2}\Big[|00\rangle + \omega^2\,|01\rangle + \omega^4\,|10\rangle + \omega^6\,|11\rangle\Big] \qquad (x_2 = 1, \text{all others} = 0),$$

$$|11\rangle \xrightarrow{\text{QFT}} \frac{1}{2}\Big[|00\rangle + \omega^3\,|01\rangle + \omega^6\,|10\rangle + \omega^9\,|11\rangle\Big] \qquad (x_3 = 1, \text{all others} = 0).$$

For all superpositions of the basis states, the final state produced by the QFT can be obtained with these formulas.

A rather remarkable fact is not readily apparent from these explicit formulas for the QFT-transformed basis states: The two qubits are not entangled in any of the states. The states can be written in the form $|\psi_1\rangle_1|\psi_2\rangle_2$, that is, as product states in which the states of the two qubits appear as separate factors. This is an important result that can be generalized to the case of n qubits. Indeed, in general, the following identity holds (which is stated here without proof and verified below only by an example):

$$\frac{1}{\sqrt{2^n}} \sum_{m=0}^{2^n-1} e^{\frac{2\pi i}{2^n} jm} |m\rangle = \frac{1}{\sqrt{2^n}} \bigotimes_{k=0}^{n-1}\Big(|0\rangle_k + e^{\frac{i\pi}{2^k}j}|1\rangle_k\Big). \qquad (6.74)$$

For each state $|j\rangle$ of the computational basis with $|j\rangle = |j_{n-1}\ldots j_2 j_1 j_0\rangle$, only one of the terms in Eq. (6.69) is different from zero, so that Eq. (6.68) yields the following formula for the transformation behavior under the QFT:

Product representation of the quantum Fourier transform:

$$|j\rangle \xrightarrow{\text{QFT}} |y\rangle = \frac{1}{\sqrt{N}} \bigotimes_{k=0}^{n-1}\Big(|0\rangle_k + e^{\frac{i\pi}{2^k}j}|1\rangle_k\Big). \qquad (6.75)$$

Again, j in the exponent is the decimal representation of the bit pattern in the initial state $|j\rangle$. The right-hand side of Eq. (6.75) is a (tensor) product of n factors, each of which describes the state of a single qubit. This becomes even more obvious when we write out the product explicitly:

$$|y\rangle = \frac{1}{\sqrt{N}}\Big(|0\rangle_0 + e^{i\pi j}|1\rangle_0\Big) \otimes \Big(|0\rangle_1 + e^{\frac{i\pi}{2}j}|1\rangle_1\Big) \otimes \cdots \otimes \Big(|0\rangle_{n-1} + e^{\frac{i\pi}{2^{n-1}}j}|1\rangle_{n-1}\Big). \qquad (6.76)$$

In the following, we suppress the tensor product notation with $\otimes$ and use indices to indicate which qubit is referred to.

Example: Using the state $|10\rangle$, verify that the formula (6.76) gives the same result as the one found in the previous example.

Solution: For two qubits, only the first two terms in Eq. (6.75) are relevant. The state $|j\rangle = |j_1 j_0\rangle = |10\rangle$ corresponds to $j = 2$ (decimal representation). So:

$$|y\rangle = \frac{1}{2}\left(|0\rangle_0 + e^{i\pi \times 2}|1\rangle_0\right) \times \left(|0\rangle_1 + e^{i\pi}|1\rangle_1\right). \tag{6.77}$$

Expanding the terms results in:

$$|y\rangle = \frac{1}{2}\left(|0\rangle_0|0\rangle_1 + e^{i\pi}|0\rangle_0|1\rangle_1 + e^{2\pi i}|1\rangle_0|0\rangle_1 + e^{3\pi i}|1\rangle_0|1\rangle_1\right). \tag{6.78}$$

Taking $\omega = e^{\frac{i\pi}{2}}$ into account, the agreement with the previous example becomes apparent (the third line of the formula there).

Implementation of the QFT by quantum gates

The product representation is the starting point for implementing the quantum Fourier transform through quantum gates. First, of course, because it shows how the QFT can be represented by operations that act on only a single qubit at a time. Second, it shows that these operations have a relatively simple structure: a phase shift of the state component $|1\rangle$, while component $|0\rangle$ remains unchanged. For each of the n qubits, the phase shift is slightly different, depending on its position in the register and the initial state j: For the initial state $|j\rangle$, the $|1\rangle$ component of qubit k is rotated by the angle $j\pi/2^k$.

The third observation is crucial for the implementation of the QFT. While it is true that the phase shift for a given qubit depends on the number j and thus on the state of the other qubits $j_{n-1}\ldots j_0$, a procedure can be given to successively work through these mutual dependencies. This is important because in the QFT, the final state results from the initial state by a unitary operation. Both states are represented by the same qubits (at different times). The qubits in which the initial state was stored are changed as the algorithm runs. Therefore, the initial state is only partially available for readout in later stages of the algorithm.

---

i To illustrate the procedure, we consider again the example of the QFT with two qubits. We write the number j in binary notation by specifying the individual bits: $j = 2^1 \times j_1 + 2^0 \times j_0$. Again, we consider only states of the computational basis so that j_1 and j_0 have well-defined numerical values 0 or 1. According to Eq. (6.75) the final state is:

$$|y\rangle = \frac{1}{2}\left(|0\rangle_0 + \underbrace{e^{2\pi i j_1}}_{=1} e^{i\pi j_0}|1\rangle_0\right) \times \left(|0\rangle_1 + e^{i\pi j_1} e^{i\frac{\pi}{2}j_0}|1\rangle_1\right). \tag{6.79}$$

We first note that $e^{2\pi i} = 1$. The same is true for all integer multiples of the exponent so that the term marked in the formula above is equal to 1. Therefore, the final state of the less significant qubit 0 does not depend on the state of the more significant qubit 1. The reverse is not true.

The behavior found here recurs when we generalize to multiple qubits: The final state of qubit k is independent of the initial states of the more significant qubits, but it depends on the less significant ones. Therefore, when running the QFT, the processing starts with the most significant qubits because if their state changes during the course of the algorithm, the subsequent processing of the less significant qubits will not be affected.

phase gate R_k in ──$\boxed{R_k}$── out

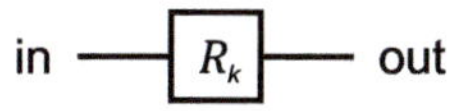

in	out
$|0\rangle$	$|0\rangle$
$|1\rangle$	$\exp(\frac{2\pi i}{2^k})\,|1\rangle$

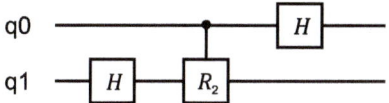

CR_k in1 ──●── out1
         in2 ──$\boxed{R_k}$── out2

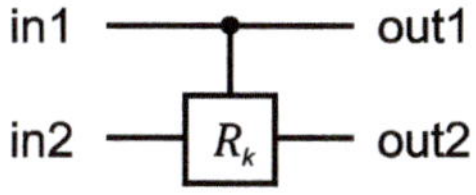

in1	in2	out1	out2
$|0\rangle$	$|0\rangle$	$|0\rangle$	$|0\rangle$
$|0\rangle$	$|1\rangle$	$|0\rangle$	$|1\rangle$
$|1\rangle$	$|0\rangle$	$|1\rangle$	$|0\rangle$
$|1\rangle$	$|1\rangle$	$|1\rangle$	$\exp(\frac{2\pi i}{2^k})\,|1\rangle$

Figure 6.20: Truth table for the R_k phase gate and its controlled variant.

q0 ──────●──$\boxed{H}$──
q1 ─$\boxed{H}$─$\boxed{R_2}$────

Figure 6.21: Realization of QFT for two qubits with quantum gates.

With these preliminary considerations, two types of quantum gates are sufficient to implement the QFT algorithm: first, the Hadamard gate, and second, a gate that performs a phase shift on the $|1\rangle$ component of one qubit depending on the state of a second qubit. The R_k gate and its controlled variant are available for this purpose. Their operation is explained in Fig. 6.20. In matrix notation, R_k has the following form:

$$R_k = \begin{pmatrix} 1 & 0 \\ 0 & \exp(\frac{2\pi i}{2^k}) \end{pmatrix}. \tag{6.80}$$

Example: Show that QFT for two qubits can be realized with the gate sequence shown in Fig. 6.21.

Solution: We go through the sequence of steps from left to right, constructing the state after applying the appropriate gate in each case. Again, we consider only the basis states of the computational basis, such that j_1 and j_0 are well-defined, predetermined numbers 0 or 1. The initial state of the system is $|\psi_0\rangle = |j_0\rangle_0|j_1\rangle_1$.

1. *Step 1: Hadamard gate on qubit 1*
   As announced, the algorithm starts by processing the most significant qubit, here, qubit 1. The state $|\psi_1\rangle = H_{(1)}|\psi_0\rangle$ generated by applying the Hadamard gate depends on whether the qubit was in state $|0\rangle$ or $|1\rangle$ at the beginning. So, we distinguish two cases:

$$j_1 = 0: \quad H_{(1)}|\psi_0\rangle = \frac{1}{\sqrt{2}}|j_0\rangle_0\big(|0\rangle_1 + |1\rangle_1\big),$$

$$j_1 = 1: \quad H_{(1)}|\psi_0\rangle = \frac{1}{\sqrt{2}}|j_0\rangle_0\big(|0\rangle_1 - |1\rangle_1\big) = \frac{1}{\sqrt{2}}|j_0\rangle_0\big(|0\rangle_1 + e^{i\pi}|1\rangle_1\big).$$

The two expressions can be combined by explicitly referring to j_1 in the exponent:

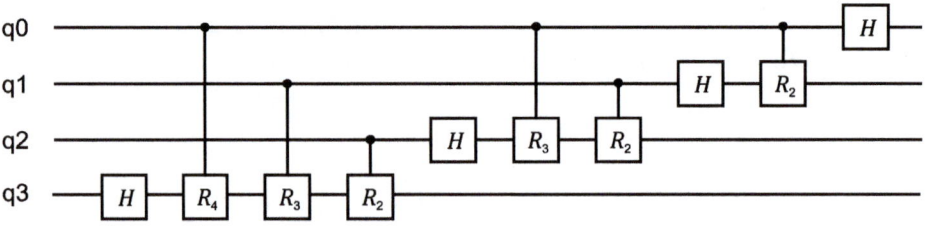

Figure 6.22: QFT for four qubits.

$$|\psi_1\rangle = H_{(1)}|\psi_0\rangle = \frac{1}{\sqrt{2}}|j_0\rangle_0\left(|0\rangle_1 + e^{i\pi j_1}|1\rangle_1\right). \tag{6.81}$$

Note that we have already obtained one of the terms in Eq. (6.79).

2. *Step 2: Controlled-R_2-gate on qubit 1*
   To generate the rest of the state of qubit 1, we need to perform a phase shift dependent on j_0. The $|1\rangle$ component should receive a phase factor $e^{i\frac{\pi}{2}}$, provided $j_0 = 1$. For $j = 0$ the state remains unchanged. This is exactly the effect of the CR_2 gate controlled by Qubit 0.:

$$|\psi_2\rangle = CR_{2(1,0)}H_{(1)}|\psi_0\rangle = \frac{1}{\sqrt{2}}|j_0\rangle_0\left(|0\rangle_1 + e^{i\pi j_1}e^{i\frac{\pi}{2}j_0}|1\rangle_1\right). \tag{6.82}$$

We have thus successfully generated the QFT-transformed state of qubit 1 in accordance with Eq. (6.79).

3. *Step 3: Hadamard gate on qubit 0*
   Now the phase shift of qubit 0 has to be produced. This is done in the same way as in step 1 by the Hadamard gate:

$$|\psi_3\rangle = H_{(0)}CR_{2(1,0)}H_{(1)}|\psi_0\rangle$$
$$= \frac{1}{2}\left(|0\rangle_0 + e^{i\pi j_0}|1\rangle_0\right) \times \left(|0\rangle_1 + e^{i\pi j_1}e^{i\frac{\pi}{2}j_0}|1\rangle_1\right). \tag{6.83}$$

To perform the last step, it was crucial that the transformation for qubit 0 does not depend on j_1, i. e., on the initial state of the more significant qubit 1, which has already been changed in the previous steps.

Fig. 6.22 shows the implementation of the QFT for four qubits. It is easy to see how it generalizes to n qubits. There are different conventions for the significance of the qubits (ascending or descending). In some cases, it, therefore, may be necessary to change their order. This can be done with a number of swap gates, which are unproblematic to implement.

From the scheme that can be inferred from the figure, we see that $\frac{1}{2}n(n-1)$ quantum gates are needed for n qubits, so the computational cost of the QFT scales with $\mathcal{O}(n^2) = \mathcal{O}((\log_2 N)^2)$. This is an exponential speedup even compared to the Fast Fourier Transform – though, as mentioned, with a restricted range of use cases.

i In the product representation of the QFT in Eq. (6.75), we see from the factor 2^k in the denominator of the exponent that the phase rotation of the qubits becomes smaller and smaller with increasing k. Qubits with large k (i. e., the most significant ones in the binary representation) are thus hardly affected by QFT, only by an increasingly small phase rotation.

The concept of *approximate QFT* is to compare the error caused by omitting these small phase rotations with the error caused by the gate operations necessary to realize them. These errors introduced by noisy gates are a strong limiting factor in today's quantum computers. Thus, as k increases, it is checked whether the error introduced by the gate operations is greater than the gain in accuracy, and the corresponding gates are omitted. In this way, one can get to the order of $\mathcal{O}(\log_2 N)$ quantum gates without significantly increasing the error already present due to noise.

6.6 Factorization of large numbers: Shor algorithm

Relevance of the Shor algorithm for computer science

The moment when the concept of quantum computing went from being a remote specialist topic to a serious field of research can be identified quite precisely. In the summer of 1994, news spread that Peter Shor [80] had specified an algorithm that could factor very large numbers on a quantum computer. At that time, there were not even rudimentarily realistic concepts for the physical realization of a quantum computer. Within a very short period of time, a new field of research emerged to study in detail the theoretical and experimental foundations of quantum computing. Why did this news, which came out of the blue at the time, cause such a stir? The answer lies in the relevance of the problem solved by the Shor algorithm.

The *factorization of large numbers* is one of the NP class problems in the complexity theory of theoretical computer science. As the number of digits increases, factoring becomes increasingly difficult for a classical computer. The computation time for these problems grows faster than polynomially with the number of digits. The inverse problem, on the other hand, is very simple: In order to verify that the large number is indeed the product of the given factors, all you have to do is multiply the factors. Now, if Shor's algorithm does indeed provide a method for solving an NP problem in polynomial time, then this is a whole new aspect that quantum computing brings to complexity theory.

Decrypting messages using the Shor algorithm

Even more important than the theoretical advances are the practical consequences of Shor's algorithm. In fact, given a quantum computer with sufficient processing power, the Shor algorithm can be used to break a number of encryption schemes previously thought to be unbreakable. Many widely used encryption schemes, such as the RSA scheme, rely on the fact that factorization belongs to the NP class and is thus a difficult problem for classical computers.

The RSA method uses a public key and a private key for each participant. A user's public key is used to encrypt messages to that recipient. The private key is used for decryption and must be kept secret (Fig. 6.23). Both keys are large numbers derived in different ways from the same product of prime numbers. The security of the method is based on the fact that factoring this product is a difficult problem. If an attacker knows

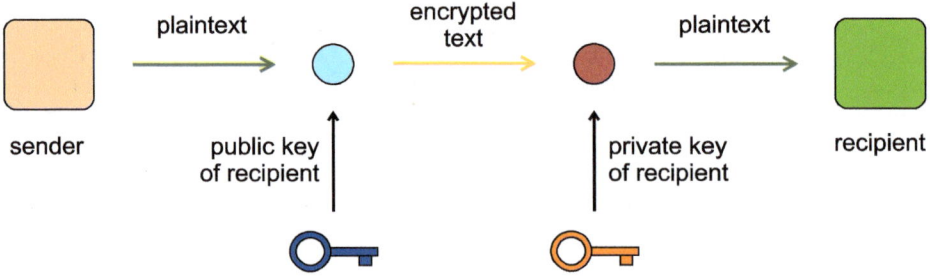

Figure 6.23: Principle of the RSA procedure.

the product representation of the public key, he can reconstruct the private key from it and decrypt encrypted messages to the recipient.

It is practically impossible for current computers to factor in sufficiently large products of primes. The largest product of two primes that has been factorized, a number known as RSA-250, has 250 digits.

Now, if the Shor algorithm significantly reduces the complexity of factoring large numbers (e. g., the products of primes mentioned above), the consequence is that encryption schemes like RSA suddenly become insecure. This has far-reaching consequences because it creates an asymmetric situation: anyone with a working quantum computer can decrypt the secret communications of other parties, while the reverse is not possible. This is a major motivation, especially for states and governments, to push the development of quantum computers.

Factorization as a search for periodicities

At first glance, it is not at all obvious how quantum mechanics could help with the factorization of large numbers. Hence, the surprise with which the Shor algorithm was initially received. David Mermin, an outstanding expert in quantum mechanics, expressed this vividly in an editorial [81]:

> But what on earth can quantum mechanics have to do with factoring? This question bothered me for four years, from the time I heard about the discovery that a quantum computer was spectacularly good at factoring until I finally took the trouble to find out how it was done. The answer, you will be relieved – but, if you're like me, also a little disappointed – to learn, is that quantum mechanics has nothing at all directly to do with factoring. But it does have a lot to do with waves. Many important waves are periodic, so it is not very surprising that quantum mechanics might be useful in efficiently revealing features associated with periodicity.

Mermin made the point: Quantum mechanics has nothing to do with factoring. However, as a wave theory, it has something to do with periodicities and can help to find periodicities. We already know the tool for this: It is the quantum Fourier transform.

Shor's algorithm works by reducing the problem of factoring, using purely classical mathematics, and finding periodicities in number sequences. These periodicities are then found by means of the quantum Fourier transform. This is the shortest description that can be given of how Shor's algorithm works. In the following, we will describe the procedure in more detail.

Factorization and modular arithmetic

The method used in the Shor algorithm uses mathematical methods from the field of modular arithmetic. Basically, this is the division of integers with a remainder. For example, dividing 15 by 6 leaves a remainder of 3:

$$15 \div 6 = 2 \text{ with remainder } 3$$

because $15 = (2 \times 6) + 3$. The modulo function simply returns the remainder of a division. So, you write:

$$15 \bmod 6 = 3. \tag{6.84}$$

In Shor's algorithm, modular arithmetic is used to find a divisor of a given number M, which is typically very large. It does not have to be a prime factor. The hard problem is finding a nontrivial divisor (not equal to 1 or M) at all. The procedure can then be continued iteratively.

To explain the procedure, we assume that we know two natural numbers a and p for which the following holds:

$$a^p \bmod M = 1. \tag{6.85}$$

This is equivalent to

$$(a^p - 1) \bmod M = 0. \tag{6.86}$$

This can be written:

$$(a^{\frac{p}{2}} + 1)(a^{\frac{p}{2}} - 1) = m \times M, \tag{6.87}$$

with integer m. Thus, since $a^p - 1$ can be divided by M without a remainder, the left side of the equation is an integer multiple of M. So there is a good chance that at least one of the two factors $(a^{\frac{p}{2}} + 1)$ or $(a^{\frac{p}{2}} - 1)$ has a nontrivial factor in common with M.

To remove the common integer multiples on both sides of Eq. (6.87), one uses the well-known and efficient Euclidean algorithm to determine the greatest common divisor (gcd). If the resulting factors are nontrivial, we are a big step closer to factorizing M.

Example: Find the prime factors of the number 15 using the procedure described with $a = 7$ and $p = 4$.

Solution: We first check if Eq. (6.85) is satisfied with the given values of a and p:

$$a^p \bmod M = 7^4 \bmod 15 = 2401 \bmod 15 = 1.$$

Since this is the case, according to Eq. (6.87), the following holds:

$$\left(a^{\frac{p}{2}} + 1\right)\left(a^{\frac{p}{2}} - 1\right) = \left(7^2 + 1\right)\left(7^2 - 1\right) = 2400 = m \times 15$$

with $m = 160$. Explicitly written:

$$50 \times 48 = 160 \times 15.$$

We now search for the factors of 15 by finding the greatest common divisor:

$$\gcd(50, 15) = 5, \quad \gcd(48, 15) = 3. \tag{6.88}$$

Thus, using the method described, we have found the prime factorization $15 = 3 \times 5$.

It seems that we have found an effective but extremely complicated method for factorizing the large number M. The method has the additional disadvantage that one has to know numbers a and p, which satisfy Eq. (6.85) to be able to use it at all. This is indeed the heart of the problem that we need to address in the following.

In addition, there seems to be another complication in the form of the Euclidean algorithm, which adds yet another layer of complexity. This concern is not justified, however, because it is an efficient algorithm whose computational complexity scales as $\mathcal{O}((\log_2 N)^3)$ even on classical computers.

Factorization and order finding
Now we come to the core of the problem: To determine natural numbers a and p for which

$$a^p \bmod M = 1. \tag{6.89}$$

In number theory, the smallest number p for which this equation holds for given a and M is called the *order* of a modulo M. Accordingly, our previous considerations have reduced the factoring of a number to find its order.

A crucial fact becomes apparent if we consider p in this equation as a variable and look at the properties of the function

$$f(p) = a^p \bmod M \tag{6.90}$$

for different values of p. An example with $a = 7$ and $M = 15$ is shown in Fig. 6.24. You immediately notice that $f(p)$ is periodic in p with period 4. The value of the period is just the order: $p = 4$ is the smallest value for which $a^p \bmod M = 1$. This relationship between periodicity and order does not just happen to hold for this particular example

period of $f(p) = 7^p$ mod 15

$7^1 = 7$	$\Rightarrow$ 7^1 mod 15 = **7**
$7^2 = 49$	$\Rightarrow$ 7^2 mod 15 = **4**
$7^3 = 343$	$\Rightarrow$ 7^3 mod 15 = **13**
$7^4 = 2401$	$\Rightarrow$ 7^4 mod 15 = **1**
$7^5 = 16{,}807$	$\Rightarrow$ 7^5 mod 15 = **7**
$7^6 = 117{,}649$	$\Rightarrow$ 7^6 mod 15 = **4**
$7^7 = 823{,}543$	$\Rightarrow$ 7^7 mod 15 = **13**
$7^8 = 5{,}764{,}801$	$\Rightarrow$ 7^8 mod 15 = **1**
$7^9 = 40{,}353{,}607$	$\Rightarrow$ 7^9 mod 15 = **7**

Figure 6.24: Example for the periodicity of a function of the form $f(p) = a^p$ mod M.

but can be used in general to determine the order (for a more detailed number theoretic background, cf., e. g., [76]).

Thus, we have achieved the goal that Mermin stated above: We have traced the factorization of large numbers back to finding periodicities. To factorize the number M, we need to determine the period of the function $f(p)$ defined in Eq. (6.90). This is a task where a quantum computer can be useful. We have already found the tool for doing this: the quantum Fourier transform. In the Shor algorithm, the various ingredients are brought together to construct a factorization algorithm that can solve the problem exponentially faster than classical algorithms.

> The *Shor algorithm* reduces the factorization of large numbers to the search for periodicities. As it uses the quantum Fourier transform, it is considerably faster than classical algorithms.

The procedure does not work without trial and error. This concerns the number a that one has to choose at the beginning. For this value of a, one tries to find the order modulo M by determining the period of $f(p)$. It may happen that the resultant p is odd. In this case, $\frac{p}{2}$ is not an integer and Eq. (6.87) does not help. Then the attempt has failed, and we have to start over with another value of a.

There are a number of other dead ends, for example, when trivial factors like M or 1 result. Problems also arise if M is a prime power. However, all these cases can be handled by classical methods [76]. It can be shown that the probability of successfully determining a nontrivial factor of M with a randomly chosen value of a is greater than 50 % (more precisely: The probability of finding a nontrivial factor of M with a randomly chosen a is greater than $1 - (\frac{1}{2})^{k-1}$, where k is the number of distinct odd prime factors of M [82]).

Putting it all together: Shor's algorithm

The above description of the factorization method did not use any quantum mechanics. It could easily be implemented on a classical computer. The period of a^p mod M would be determined by a classical FFT. However, such an implementation would have no advantage over other classical factorization methods.

The dramatic speed-up in Shor's algorithm is due to the quantum Fourier transform, which replaces the classical FFT. It is, in fact, the only piece of quantum mechanics involved in the Shor algorithm.[1] Indeed, period finding is an ideal application for QFT because periodicity is a global feature where quantum parallelism can be fully exploited. Recall that although all Fourier components are contained in the final state of QFT, we cannot read them out simultaneously in a measurement. The measurement returns only a single value. For period finding, however, this single value may be sufficient. Ideally, therefore, Shor's algorithm can be run with a single call of the QFT – and as we have seen, this means an exponential speedup over the classical FFT.

We will now go through Shor's algorithm step by step. We follow the procedure in Shor's paper of 1999 [82]. We look for a factor of the number M by determining the period of the function

$$f(p) = a^p \bmod M. \tag{6.91}$$

To make the representation less abstract, we stay with the concrete example considered above and factorize the number $M = 15$. As an arbitrary value of a, we choose $a = 7$.

1. *Step 1: Preparation*
   We check if the randomly chosen number a is not a divisor of M: $\gcd(a, M) = 1$. Otherwise, we have found a divisor of M and are done.

2. *Step 2: Determine the number of qubits needed*
   The Shor algorithm uses two quantum registers. The first register contains the binary representation of the numbers from 0 to $q - 1$, i. e. the argument p of the function $f(p)$. The second register is occupied by the function value $f(p)$ during the calculation: the binary representation of $a^p \bmod M$ (Fig. 6.25).
   In the algorithm, the number q is generally chosen to be a power of two with $M^2 \leq q \leq 2M^2$. In our case, this would be $q = 256 = 2^8$. So, $2 \times 8 = 16$ qubits would be needed. However, it turns out that for our particular example, the numerical values are such that $q = 2^4 = 16$ will also work. This corresponds to $2 \times 4 = 8$ qubits. Since the formulas we are going to write down will have significantly fewer terms with this choice, we will take advantage of it. For better readability, the explicit value $q = 16$ is written in the formulas. In the general case, other values of q have to be inserted accordingly.

3. *Step 3: Initialization of the first register*
   Starting with all qubits in the state $|0\rangle$, the first quantum register is brought into the symmetric superposition state. This is done by applying the Hadamard gate to all qubits of the first register. The resulting state is:

---

[1] Which shows, by the way, that it is not quantum mechanics that makes the Shor algorithm so difficult for many to understand, but the unfamiliar mathematics (modular arithmetic and number theory).

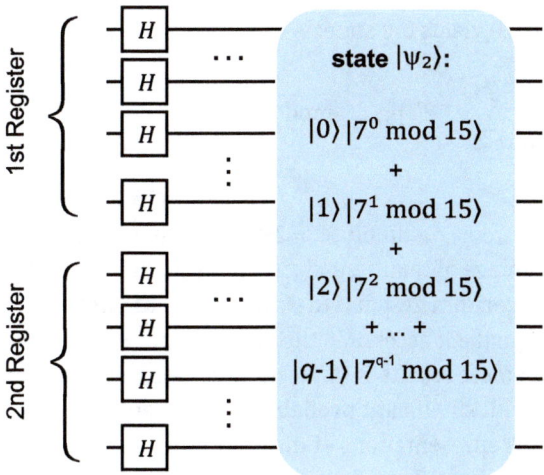

Figure 6.25: Quantum registers and state $|\psi_2\rangle$ in the Shor algorithm.

$$|\psi_1\rangle = \frac{1}{\sqrt{16}} \sum_{p=0}^{15} |p\rangle |0\rangle . \tag{6.92}$$

4. *Step 4: Calculate $a^p \bmod M$ in the second register*
    To find the period of the function $f(p) = a^p \bmod M$, the second register is assigned the function value $f(p)$, where p is the number encoded in the first register. The resulting state is:

$$|\psi_2\rangle = \frac{1}{\sqrt{16}} \sum_{p=0}^{15} |p\rangle |7^p \bmod 15\rangle . \tag{6.93}$$

Written out with concrete numbers, the state is:

$$|\psi_2\rangle = \frac{1}{\sqrt{16}} [|0\rangle |1\rangle + |1\rangle |7\rangle + |2\rangle |4\rangle + |3\rangle |13\rangle + |4\rangle |1\rangle + |5\rangle |7\rangle$$
$$+ |6\rangle |4\rangle + |7\rangle |13\rangle + \cdots + |14\rangle |4\rangle + |15\rangle |13\rangle] . \tag{6.94}$$

In the second quantum register, the recurring sequence of numbers $1, 7, 4, 13$ reflects the periodicity already seen in Fig. 6.24, which we will now detect mathematically using the QFT.

5. *Step 5: Apply QFT to the first register*
    As shown in Eq. (6.68), the QFT has the following effect on each of the basis states $|p\rangle$:

$$|p\rangle \longrightarrow \frac{1}{\sqrt{q}} \sum_{k=0}^{q-1} e^{2\pi i \frac{pk}{q}} |k\rangle . \tag{6.95}$$

Thus, for our concrete case, Eq. (6.93) yields the state:

$$|\psi_3\rangle = \frac{1}{16} \sum_{p=0}^{15} \sum_{k=0}^{15} e^{\frac{2\pi i}{16} p\,k} |k\rangle \, |7^p \bmod 15\rangle . \tag{6.96}$$

6. *Step 6: Measurement*

In the next step, all qubits are measured. Potentially $2^8 = 256$ different measurement results are possible – it is the advantage of quantum algorithms that the state space is very large. The art of quantum algorithm design is to extract relevant information from these many possibilities and make it accessible through measurement. In the Shor algorithm, the probability of most results is (almost) zero. Only a few values will be found in a measurement with significant probability. For example, we can see from Eq. (6.94) that only the bit representations of the numbers 1, 4, 7, or 13 will be found in the second register because others do not occur in the state. Each of them has a probability of $\frac{1}{4}$.

Let us assume that the measurement yields the value 13 for the second register. We explicitly write out only those parts of $|\psi_3\rangle$ that are relevant to this measurement result. From Eq. (6.94), we see that the number p can only take the values 3, 7, 11, 15, and so:

$$|\psi_3\rangle = \frac{1}{16} \sum_{k=0}^{15} [\underbrace{e^{\frac{2\pi i}{16} 3k}}_{p=3} + \underbrace{e^{\frac{2\pi i}{16} 7k}}_{p=7} + \underbrace{e^{\frac{2\pi i}{16} 11k}}_{p=11} + \underbrace{e^{\frac{2\pi i}{16} 15k}}_{p=15}] |k\rangle \, |13\rangle + \text{irrelevant terms.} \tag{6.97}$$

The probability of measuring a value n in the first register and a value of 13 in the second register is obtained by squaring

$$P(n,13) = |\langle n,13|\psi_3\rangle|^2 = \left| \frac{1}{16} \sum_{k=0}^{15} [\text{terms as above}] \underbrace{\langle n|k\rangle}_{=\delta_{nk}} \underbrace{\langle 13|13\rangle}_{=1} \right|^2 . \tag{6.98}$$

Evaluating the terms, we get:

$$P(n,13) = \left| \frac{1}{16} \left[\sum_{j=0}^{3} e^{\frac{2\pi i}{16} n \times j \times 4} \right] e^{\frac{2\pi i}{16} 3n} \right|^2 \tag{6.99}$$

The sum in square brackets can be further simplified. It corresponds to the orthogonality relation Eq. (6.62) with $M = 4$. Consequently, the sum is only different from zero if n is an integer multiple of 4:

$$\sum_{j=0}^{3} e^{\frac{2\pi i}{16} n \times j \times 4} = \begin{cases} 4 & \text{if } n \text{ an integer multiple of 4,} \\ 0 & \text{else.} \end{cases} \tag{6.100}$$

So, in total:

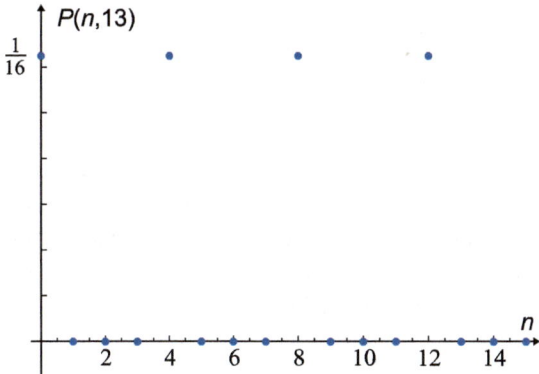

Figure 6.26: The probability $P(n, 13)$ is different from zero only if n is equal to an integer multiple of the period $p = 4$.

$$P(n, 13) = \begin{cases} \frac{1}{16} & \text{for } n = 0, 4, 8, 12, \\ 0 & \text{else.} \end{cases} \tag{6.101}$$

Thus, the measurement of n always yields a multiple of 4 (Fig. 6.26). An analogous calculation shows that this is also true if the measurement result for the second quantum register is not 13 but one of the other possible values.

7. *Step 7: Evaluation*

What was said on p. 165 of about the discrete Fourier transform also applies to the Shor algorithm: The maxima of the DFT (here: the most probable measurement results after the QFT) are integer multiples of q/p, (q = length of the analyzed sequence, p = searched period), so in our example, multiples of $16/4 = 4$. The last step of the algorithm is to derive the period p from the multiple of q/p found in the measurement. This can be done by trial and error or more systematically using methods from number theory.

Suppose the measurement yielded the value 12. Then we know that p has to be an integer multiple of $\frac{16}{12} = \frac{4}{3}$ and quickly arrive at the correct value $p = 4$. If we get 8 as the measured value, p must be an integer multiple of 2. Only if we get the value 0, the run of the algorithm has failed and needs to be repeated. Once that we have obtained the correct value $p = 4$ for the period, i. e. for the order of 7^p mod 15, we proceed as in the example on p. 177 and obtain the numbers 3 and 5 as factors of 15.

Comments on the Shor algorithm

1. To illustrate the Shor algorithm, we have deliberately chosen an example that works particularly "smoothly". The length of the sequence ($q = 16$) is just an integer multiple of the period ($p = 4$). In general, this will not be the case. Then the leakage effect discussed on p. 166 occurs: In the measurement, not only the correct period is found, but with a certain probability, also neighboring values. Therefore, classical

"post-processing" is usually necessary. Continued fraction methods are used to find suitable "candidates", and often several of these candidates have to be tried. None of these difficulties has anything to do with quantum physics: As we have seen, the leakage effect already occurs in the classical discrete Fourier transform.

2. The need for classical post-processing shows that classical and quantum computers will need to work closely together for many applications. Hybrid solutions with both classical and quantum computers are therefore useful. They will become important for practical applications. Since computing with quantum computers is expensive, it is worthwhile to assign only those tasks to them where they offer a real advantage.

3. What is the first quantum register actually used for? The periodicity already shows up in Eq. (6.94) if only the second register is considered. The answer is again related to the need for reversibility in quantum algorithms. The function $f(p) = a^p$ mod M cannot be implemented by reversible quantum operations on a single register alone because it is not injective: you cannot uniquely infer p from $f(p)$. There are several ways to make the function $f(p)$ reversible using auxiliary registers. Shor chooses the simplest way and keeps the original value p in the first register.

4. With the function $f(p) = a^p$ mod M, there is a second, more practical difficulty. Implementing $f(p)$ with quantum gates, while possible in principle, is quite cumbersome. So far, no one has come up with a better way to do this than to emulate the corresponding classical algorithms in the quantum domain. Since reversibility must be preserved, this is relatively cumbersome. Therefore, the computation of $f(p)$ is the bottleneck that limits the speed of the Shor algorithm. The effort required for $f(p)$ is also the reason why we will not discuss how to implement the Shor algorithm using quantum gates.

6.7 Quantum phase estimation

Quantum phase estimation (QPE) [83, 22] is an important component of many quantum algorithms. It is not a standalone quantum algorithm in the strict sense but rather a useful subroutine. The aim is to return the eigenvalue of a unitary operator U as a binary number if the input is one of the eigenstates of U.

The properties of unitary operators have already been discussed in Section 3.5. Their eigenvalues are complex and have modulus 1. Therefore, they can be expressed in the form

$$\lambda = e^{2\pi i \theta}. \tag{6.102}$$

The number θ is called *phase* and can take values between 0 and 1. The goal of quantum phase estimation is to read out the value of θ belonging to an eigenvector $|u\rangle$ for a given operator U and write it in binary form into a quantum register.

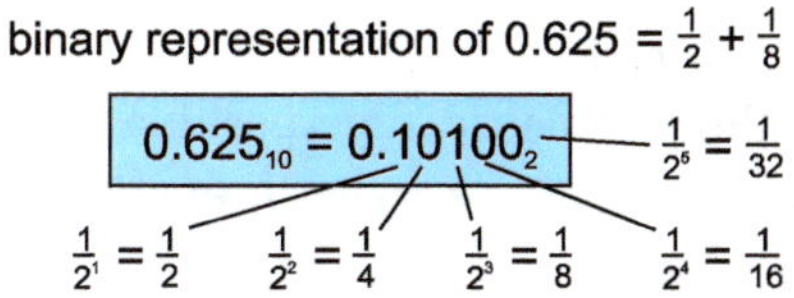

binary representation of $0.625 = \frac{1}{2} + \frac{1}{8}$

$$0.625_{10} = 0.10100_2 \quad\quad \frac{1}{2^5} = \frac{1}{32}$$

$$\frac{1}{2^1} = \frac{1}{2} \quad \frac{1}{2^2} = \frac{1}{4} \quad \frac{1}{2^3} = \frac{1}{8} \quad \frac{1}{2^4} = \frac{1}{16}$$

Figure 6.27: Binary representation of a decimal fraction.

As a reminder, unitary operators are defined by the property $U^\dagger U = \mathbb{1}$. If $|u\rangle$ is an eigenvector of U and λ is the associated eigenvalue, then

$$1 = \langle u|u\rangle = \langle u|\mathbb{1}|u\rangle = \langle u|U^\dagger U|u\rangle = |\lambda|^2 \langle u|u\rangle = |\lambda|^2. \tag{6.103}$$

This shows that λ is a complex number of modulus 1, i. e., of the form given in Eq. (6.102). Thus, the eigenvalue equation is $U|u\rangle = e^{2\pi i\theta}|u\rangle$.

The value of θ is a number between 0 and 1, which we usually write as a decimal fraction, e. g. $\theta = 0.625$. In quantum phase estimation, however, the result is returned as a binary fraction. Fig. 6.27 shows the systematics behind this: The decimal places correspond to powers of two with negative exponents. For example, the number 0.625 can be represented as $1 \times 2^{-1} + 1 \times 2^{-3}$ and therefore has the binary representation 0.101.

In the decimal system, fractions with denominators such as 3, 6, or 7 cannot be represented as finite decimal fractions. They have infinite (periodic) decimal places. In general, in arbitrary number systems, numbers have a finite representation only if their denominator is composed of the prime factors of the base of the number system. In the case of the binary system, the only prime factor is 2. Therefore, only sums of powers of two (as shown in Fig. 6.27) have finitely many decimal places. All other numbers can only be approximately represented by finite binary fractions.

Procedure for quantum phase estimation

To implement the algorithm, we need two quantum registers. The first one will eventually hold the binary representation of the number θ. It has n qubits corresponding to the desired number of binary decimal places. The number of decimal places determines the precision to which θ is approximated. The second quantum register contains the eigenstate $|u\rangle$ of the unitary operator U. It plays a mostly passive role and is only read out. As a tool, we need the quantum Fourier transform (or its inverse, which, because of $U^{-1} = U^\dagger$, is obtained simply by changing the sign in the exponent of Eq. (6.70)). Let us go through the process of quantum phase estimation step by step:

1. *Step 1: Initialization*

   In the initial state, all qubits of the first quantum register are in the state $|0\rangle$. The second quantum register contains the eigenstate $|u\rangle$ to be studied:

$$|\psi_0\rangle = |0_1 \ldots 0_n\rangle |u\rangle. \tag{6.104}$$

2. *Step 2: Hadamard gates on the first quantum register*
   The first quantum register is brought into the symmetric superposition state by applying Hadamard gates to all qubits. With $N = 2^n$, we write in analogy to Eq. (6.92):

   $$|\psi_1\rangle = \frac{1}{\sqrt{N}} \sum_{m=0}^{N-1} |m\rangle |u\rangle . \tag{6.105}$$

3. *Step 3: Apply U^m to each term of the sum*
   The operator U^m is now applied to the eigenstate $|u\rangle$ (the second quantum register) for each term of the sum. U^m means m-fold application of U. The result of this operation follows directly from the eigenvalue equation:

   $$U^m |u\rangle = e^{2\pi i\theta m} |u\rangle \tag{6.106}$$

   so that the total state is:

   $$|\psi_2\rangle = \frac{1}{\sqrt{N}} \sum_{m=0}^{N-1} |m\rangle U^m |u\rangle = \frac{1}{\sqrt{N}} \sum_{m=0}^{N-1} |m\rangle e^{2\pi i\theta m} |u\rangle . \tag{6.107}$$

   We write this as follows:

   $$|\psi_2\rangle = \frac{1}{\sqrt{N}} \sum_{m=0}^{N-1} (e^{2\pi i\theta m} |m\rangle) |u\rangle . \tag{6.108}$$

   The term in parentheses indicates that this is a phase kickback where the phase factor is transferred from $|u\rangle$ to $|m\rangle$. The state $|u\rangle$ (i. e., the second quantum register) is a common factor that we can ignore from now on. In the following, we will only write down the state of the first quantum register:

   $$|\psi_2\rangle = \frac{1}{\sqrt{N}} \sum_{m=0}^{N-1} e^{2\pi i\theta m} |m\rangle . \tag{6.109}$$

4. *Step 4: Inverse quantum Fourier transform*
   The state $|\psi_2\rangle$ has the form of Eq. (6.68) with $x_m = \frac{1}{\sqrt{N}} e^{2\pi i\theta m}$. We apply the inverse quantum Fourier transform and get:

   $$|\psi_3\rangle = \sum_{k=0}^{N-1} y_k |k\rangle \tag{6.110}$$

   with

   $$y_k = \frac{1}{N} \sum_{m=0}^{N-1} e^{2\pi i\theta m} e^{-\frac{2\pi i}{N} m k} = \frac{1}{N} \sum_{m=0}^{N-1} e^{\frac{2\pi i}{N}(N \times \theta - k)m} . \tag{6.111}$$

We will now assume that $N \times \theta$ is an integer (and explain the assumption at the end of the calculation). In this case, y_k has exactly the form of the orthogonality relation (6.62) for the complex exponential function. This means: y_k will be different from zero (for the given range of values of k and θ) only if the term in parentheses in the exponent is zero:

$$y_k = \begin{cases} 1 & \text{if } k = N \times \theta, \\ 0 & \text{else.} \end{cases} \tag{6.112}$$

This reduces the sum in Eq. (6.110) to a single term with $k = N \times \theta = 2^n \times \theta$. This gives the final state of quantum phase estimation:

$$|\psi_3\rangle = |2^n \times \theta\rangle . \tag{6.113}$$

What does it mean to assume that $N \times \theta$ is an integer, and how can we interpret the final result (6.113)? For a better understanding, let's look again at the binary representation of θ in Fig. 6.27. Multiplication by $N = 2^n$ means that all digits are shifted to the left by n places, for example, $2^3 \times 0.101_2 = 101_2$. This works in the binary system just as well as in the decimal system, where the digits are shifted to the left by multiplication with a power of ten.

Thus, the assumption of $N \times \theta$ being integer means that if the binary representation of the number θ is shifted to the left by n digits, the result is an integer without decimal places. Or, equivalently: the number θ can be represented exactly with n places in the binary system.

The result (6.113) can thus be interpreted as follows: The n qubits in the first register contain the first n decimal places of θ in binary representation. If the algorithm runs without errors, the result will be correct if θ can be represented exactly with this number of places.

Since, in reality, θ is not known in advance, we cannot know whether this assumption is true. If it is wrong, we have to expect leaking, i. e., nonzero probabilities for slightly different bit patterns. A larger number of places can reduce this error [22].

Implementation of quantum phase estimation by quantum gates

With our previous experience, implementing quantum phase estimation using quantum gates is not particularly difficult. The most complicated step is to reduce Eq. (6.107) to operations on one or two qubits. We can proceed analogously to the implementation of the quantum Fourier transform. We compare Eq. (6.109) and Eq. (6.74) and find that the terms are identical if we set $j = \theta \times 2^n$. So for $|\psi_2\rangle$ we can immediately write:

$$|\psi_2\rangle = \frac{1}{\sqrt{2^n}} \bigotimes_{k=0}^{n-1} (|0\rangle_k + e^{i\pi\theta 2^{n-k}} |1\rangle_k). \tag{6.114}$$

From this expression, we can read off how to realize $|\psi_2\rangle$ by operations on individual qubits. For each qubit k, a phase shift of the state component $|1\rangle_k$ is performed, while the state component $|0\rangle_k$ remains unchanged. Similar to QFT, the state $|\psi_2\rangle$ is realized by applying controlled U gates as shown in Fig. 6.28.

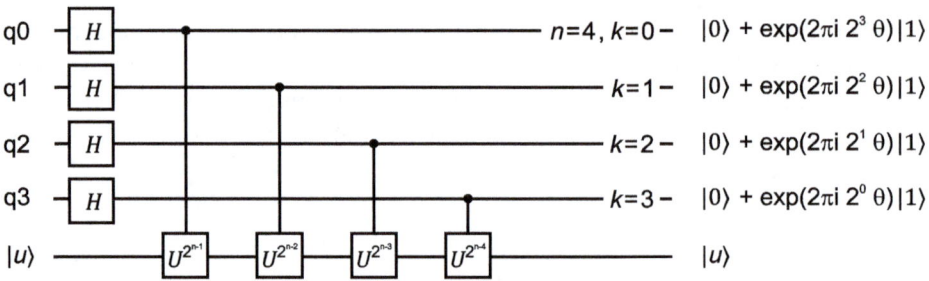

Figure 6.28: Implementation of $|\psi_2\rangle$ for $n = 4$ qubits.

 Example: Establish the connection between the formula representation Eq, (6.114) and the quantum gate representation in Fig. 6.28 by considering the effect of the operator U on the eigenstate $|u\rangle$.

Solution: We reintroduce factor $|u\rangle$ and read the eigenvalue equation (6.106) from right to left. The result is:

$$|\psi_2\rangle = \frac{1}{\sqrt{2^n}} \bigotimes_{k=0}^{n-1} \left(|0\rangle_k |u\rangle + |1\rangle_k U^{2^{n-k-1}} |u\rangle \right). \tag{6.115}$$

Thus, the operator $U^{2^{n-k-1}}$, controlled by qubit k, acts on $|u\rangle$ to generate the desired phase shift for the state component $|1\rangle_k$ via phase kickback. This results in the sequence of controlled-U operations shown in Fig. 6.28.

Example: Heads or tails detector

Let us demonstrate how quantum phase estimation works with a simple example. We take the game "Quantum Penny Flip" from Section 6.2. We construct an algorithm that is given either "heads" (i. e., $|0\rangle$) or "tails" (i. e., $|1\rangle$) as an input state $|u\rangle$. The algorithm's task is to write the number 0 or 1 into an output qubit accordingly. We need two qubits: one for the input and one for the output.

 Example: Show that $|0\rangle$ and $|1\rangle$ are eigenstates of the operator Z (one of the Pauli matrices of Eq. (3.53)). Determine the phase to be expected.

$$Z = \begin{pmatrix} 1 & 0 \\ 0 & -1 \end{pmatrix}. \tag{6.116}$$

Solution: We solve the problem by explicit calculation:

$$Z |0\rangle = \begin{pmatrix} 1 & 0 \\ 0 & -1 \end{pmatrix} \begin{pmatrix} 1 \\ 0 \end{pmatrix} = \begin{pmatrix} 1 \\ 0 \end{pmatrix} = |0\rangle. \tag{6.117}$$

The eigenvalue for $|u\rangle = |0\rangle$ is $+1$, so the phase to be detected is $\theta = 0$.

$$Z |1\rangle = \begin{pmatrix} 1 & 0 \\ 0 & -1 \end{pmatrix} \begin{pmatrix} 0 \\ 1 \end{pmatrix} = \begin{pmatrix} 0 \\ -1 \end{pmatrix} = -|1\rangle. \tag{6.118}$$

Here the eigenvalue is -1 and the phase to be detected is $\theta = \frac{1}{2}$ (because $e^{2\pi i \times \frac{1}{2}} = -1$).

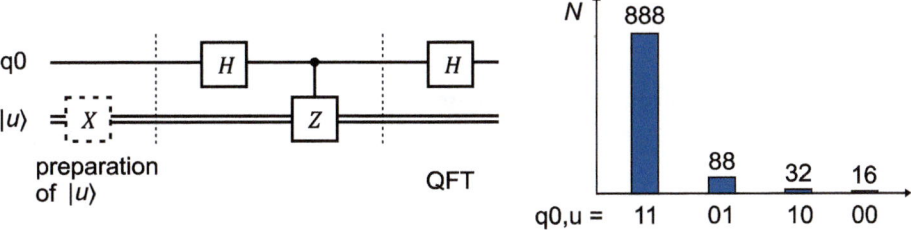

Figure 6.29: Implementation of the quantum phase estimation for the determination of heads or tails and the result of a real calculation.

With Z, we have found a unitary operator whose eigenvalues are the input states considered in the problem. Therefore, we use quantum phase estimation with $U = Z$ and $n = 1$ to determine heads or tails. As a result, we expect a value of 0 in the output bit for $|u\rangle = |0\rangle$ and 1 for $|u\rangle = |1\rangle$ (because here the phase is $\theta = \frac{1}{2}$, and in binary representation $\frac{1}{2} = 2^{-1} = 0.1_2$).

Fig. 6.29 shows the implementation of quantum phase estimation for heads-or-tails determination. Before starting the algorithm, the desired state (heads or tails) must be prepared in the $|u\rangle$ register. This is done with the optional X gate shown dashed on the left. It transforms the initial state $|0\rangle$ into $|1\rangle$ if desired. The actual quantum phase estimation is performed according to Eq. (6.115) or Fig. 6.28 with subsequent inverse QFT. The product representation reduces to a single term with $n = 1$ and $k = 0$. This implies a one-time application of the controlled Z gate. The inverse quantum Fourier transform is almost trivial. It reduces to a single Hadamard gate, so the entire quantum algorithm has the simple form shown in Fig. 6.29.

The result of a calculation on a real quantum computer is shown in Fig. 6.29 on the right. The input state is $|u\rangle = 1$, so the dashed X gate is used. The histogram shows the results of the state measurement on the two qubits for 1024 runs. Given an error-free implementation of the calculation, one would expect the result to be 11 in 100 % of the cases. In the vast majority of runs, this result is found. However, in a non-negligible number of cases, a different result is found. Current implementations of quantum computers are still so noisy that even with such a simple algorithm, significant miscalculations occur.

Example: Demonstrate that the quantum phase estimation implementation shown in Fig. 6.29 gives the desired result in the heads or tails determination.

Solution: We start with the product representation (6.115). For our case with $n = 1$, the product reduces to one term with $k = 0$:

$$|\psi_2\rangle = \frac{1}{\sqrt{2}}\left(|0\rangle\,|u\rangle + |1\rangle\,Z^{2^0}\,|u\rangle\right) = \frac{1}{\sqrt{2}}\left(|0\rangle\,|u\rangle + |1\rangle\,Z\,|u\rangle\right). \tag{6.119}$$

Using the eigenvalue equation $Z\,|u\rangle = \pm\,|u\rangle$, we obtain for $|\psi_2\rangle$:

$$|\psi_2\rangle = \frac{1}{\sqrt{2}} \big(|0\rangle \pm |1\rangle \big) |u\rangle \quad \begin{cases} \text{for } |u\rangle = |0\rangle \,, \\ \text{for } |u\rangle = |1\rangle \,. \end{cases} \tag{6.120}$$

To arrive at the final state of the algorithm, a last inverse quantum Fourier transform is necessary, which reduces to the application of the Hadamard gate (analogous to Fig. 6.21). According to Eq. (6.11), the expression for the Hadamard gate in the Dirac notation is as follows:

$$H = \frac{1}{\sqrt{2}} \big[|0\rangle \langle 0| + |1\rangle \langle 0| + |0\rangle \langle 1| - |1\rangle \langle 1| \big]. \tag{6.121}$$

Applying this operator to qubit q_0 in $|\psi_2\rangle$ yields:

$$H|\psi_2\rangle = \frac{1}{2} \big(|0\rangle + |1\rangle \pm |0\rangle \mp |1\rangle \big) |u\rangle = \begin{cases} |0\rangle |u\rangle & \text{for } |u\rangle = |0\rangle \,, \\ |1\rangle |u\rangle & \text{for } |u\rangle = |1\rangle \,. \end{cases} \tag{6.122}$$

Thus, the first quantum register (qubit q_0) contains the desired information about heads or tails in the final state.

6.8 Linear systems of equations: the HHL algorithm

One of the most important tasks in numerical mathematics is the solution of linear systems of equations. The problem is to solve a system of N equations with N unknowns. It can be formulated compactly in matrix notation: Given an $N \times N$ matrix A and a vector $\vec{b}$, the goal is to find a vector $\vec{x}$ so that the equation

$$A \times \vec{x} = \vec{b} \tag{6.123}$$

is satisfied. Formally, the solution can be written as $\vec{x} = A^{-1} \times \vec{b}$. Thus, solving a system of linear equations is equivalent to inverting the matrix A.

Many practical problems require the solution of large linear systems with many unknowns. The problem arises in a wide range of applications, from climate research to mechanical engineering, medical research, finance, and machine learning. In engineering, the finite element method, which reduces partial differential equations to linear systems, is widely used in design and modeling in many fields.

In many application problems, the resulting coefficient matrices A are sparse: Most entries are zero because, for example, only neighboring domains interact in the systems being modeled.

In classical computer science, there are powerful and well-developed methods for solving very large linear systems of equations. And yet, in 2009 Harrow, Hassidim, and Lloyd (HHL) introduced a quantum algorithm that can achieve exponential speedup for certain problems (e. g., with sparse coefficient matrices) [84].

---

i Even in classical computer science, problems can arise when solving systems of linear equations. For example, two of the equations may be linearly dependent, or one of the equations may be expressed as a linear combination of the other equations. Then the system of equations is underdetermined and is called *singular*. It is also possible that the linear dependence of the equations is not exact but only approximate. This is called an *ill-conditioned* problem. The *condition number* of a matrix is the ratio of the largest to the smallest of its eigenvalues. Ill-conditioned problems have large condition numbers. There are special numerical algorithms, such as singular value decomposition, to deal with ill-conditioned problems. The HHL algorithm has a lower probability of finding a solution for ill-conditioned problems and thus loses its advantages in this case.

For the following discussion, we assume that the matrix A is hermitian. There are ways to extend the algorithm to more general matrices. We refer to the further literature on this topic, e. g., [85, 86]. Finally, we assume that the matrix A is sparse. This is not essential for the operation of the HHL algorithm, but the superiority over classical algorithms is more pronounced in this case.

Principle of the HHL algorithm

Before we dive into the details, let us take a look at how the HHL algorithm works in general. The difference to classical algorithms starts with the encoding of the data. Instead of a classical vector $\vec{x}$ with N bits, a quantum state $|x\rangle$ with N qubits is used. The vector $\vec{b}$ on the right side is also represented by an N-qubit state $|b\rangle$. It is used as input and has to be prepared accordingly. So, the problem to solve is:

$$A\,|x\rangle = |b\rangle \quad \text{or} \quad |x\rangle = A^{-1}\,|b\rangle\,. \tag{6.124}$$

The quantum version (6.124) differs from the classical problem in various ways:

1. The state vectors $|x\rangle$ and $|b\rangle$ are always normalized. Thus, only an appropriately scaled version of the original problem can be considered.
2. The result of the algorithm is a solution state $|x\rangle$. It contains the components of the solution vector as probability amplitudes. Thus, the quantum algorithm is fundamentally different from its classical counterparts, where the solution vector is directly available as a classical bit pattern in an output register.
   Once again, we encounter here a well-known peculiarity of quantum algorithms: The solution of the problem cannot be read out directly because the amplitudes of the solution state cannot be determined by a single measurement. Thus, the HHL algorithm can be advantageous only for those problems where the state $|x\rangle$ is not needed directly but is used for other calculations, e. g., for probabilities $|\langle j|x\rangle|^2$ or expectation values $\langle x|M|x\rangle$.

The hermitian matrix A has N eigenvalues λ_j and eigenvectors $|u_j\rangle$ for which the eigenvalue equation

$$A\,|u_j\rangle = \lambda_j\,|u_j\rangle \tag{6.125}$$

holds. Thus, according to Eq. (3.61), A can be written in the following way:

$$A = \sum_{j=1}^{N} \lambda_j\,|u_j\rangle\,\langle u_j|\,. \tag{6.126}$$

In a concrete situation, we do not know λ_j and $|u_j\rangle$, and we will never know them in the course of the calculation. However, to understand the principle of the HHL algorithm, we need to express the matrix A in the representation (6.126) in which it has a diagonal form. In fact, it is quite easy to invert A in this representation. To invert a diagonal matrix, you simply take the inverse of the diagonal elements:

$$
A = \begin{pmatrix} \lambda_1 & & & \\ & \lambda_2 & & \\ & & \ddots & \\ & & & \lambda_N \end{pmatrix} \quad \Rightarrow \quad A^{-1} = \begin{pmatrix} \lambda_1^{-1} & & & \\ & \lambda_2^{-1} & & \\ & & \ddots & \\ & & & \lambda_N^{-1} \end{pmatrix}. \tag{6.127}
$$

Formally, the inverse of A is therefore

$$
A^{-1} = \sum_j \frac{1}{\lambda_j} |u_j\rangle \langle u_j| . \tag{6.128}
$$

Our problem $|x\rangle = A^{-1} |b\rangle$ can be solved by substituting A^{-1} from above:

$$
|x\rangle = \sum_j \frac{1}{\lambda_j} |u_j\rangle \langle u_j|b\rangle \tag{6.129}
$$

and with $b_j = \langle u_j|b\rangle$:

$$
|x\rangle = \sum_j \frac{b_j}{\lambda_j} |u_j\rangle . \tag{6.130}
$$

Thus, we have a formal solution, but it is useless without knowing the λ_j and $|u_j\rangle$. This is where the HHL algorithm comes in. It uses quantum phase estimation to determine the λ_j and writes them into a quantum register. They have to be numerically inverted to prepare the solution state $|x\rangle$. This is the basic idea of the algorithm. Let us now take a look at how the above ideas are put into practice.

Quantum registers

The HHL algorithm requires the following quantum registers:

1. The eigenvalues λ_j are determined by quantum phase estimation. The first quantum register records the eigenvalues in "digital", i. e., binary encoded form (Fig. 6.30). As in Eq. (6.104), the size of this register is determined by the desired number m of digits, i. e., the desired accuracy.
2. The second quantum register contains $|b\rangle$ as initial state and later $|x\rangle$ as result in "analog" form, i. e., as amplitudes in the computational basis. This requires N qubits.
3. Finally, a single ancilla qubit is needed for the algorithm.

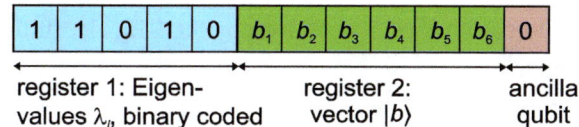

register 1: Eigen-   register 2:   ancilla
values λ_j, binary coded   vector $|b\rangle$ qubit

Figure 6.30: Quantum registers in the HHL algorithm. Entries are for illustration purposes only; the qubits are generally in superposition states.

Procedure of the HHL algorithm

We will go through the algorithm step by step:

1. *Step 1: Initialization*

   In the initial state, the first quantum register and the ancilla qubit are set to the state $|0\rangle$. In the second quantum register, the state $|b\rangle$ is prepared. This task already requires a number of quantum operations, which we will not discuss here. So, the initial state is:

   $$|\psi_0\rangle = |0\rangle\,|b\rangle\,|0\rangle_a. \tag{6.131}$$

   The ancilla qubit marked with the index a is not needed until the third step. To make the formulas clearer, we omit it for now. We write $|b\rangle = \sum b_j\,|u_j\rangle$ and thus formally express the initial state in terms of the eigenstates $|u_j\rangle$:

   $$|\psi_0\rangle = \sum_j |0\rangle\,b_j\,|u_j\rangle\,. \tag{6.132}$$

2. *Step 2: Quantum phase estimation*

   Quantum phase estimation (QPE) is used to write the eigenvalues λ_j into the first quantum register. For this we consider the operator $U = e^{iAt}$. Since we have assumed A to be hermitian, U is unitary, as required for QPE. According to the eigenvalue equation, the following holds:

   $$e^{iAt}\,|u_j\rangle = e^{i\lambda_j t}\,|u_j\rangle\,. \tag{6.133}$$

   The QPE assumes values between 0 and 1 for the phase. Therefore, we choose the constant t to shift the binary representation of λ_j by m binary places to the right (m = size of the first quantum register). With the additional factor 2π assumed in QPE, we get $t = 2\pi/2^m$. With these parameters, the QPE writes an m-bit approximation of λ_j into the first quantum register. After the QPE, the state of the system is:

   $$|\psi_1\rangle = \mathrm{QPE}(e^{iAt})\,|\psi_0\rangle = \sum_j |\lambda_j\rangle\,b_j\,|u_j\rangle\,. \tag{6.134}$$

   This state already shows some similarity to the desired final state (6.130). For each term of the sum, the value of λ_j is encoded in binary form in the first quantum register. Now the factor $\frac{1}{\lambda_j}$ has to be generated. This is done by the HHL algorithm in a somewhat brute-force manner.

3. *Step 3: Controlled rotation of the ancilla qubit*
In this step we need the ancilla qubit. It is still in state $|0\rangle_a$, so the total state is:

$$|\psi_1\rangle = \sum_j b_j \, |\lambda_j\rangle \, |u_j\rangle \, |0\rangle_a. \tag{6.135}$$

A unitary operation is now applied to the ancilla qubit controlled by the state of the first quantum register. Thus, it depends on λ_j:

$$|\psi_2\rangle = \sum_j b_j \, |\lambda_j\rangle \, |u_j\rangle \, (\alpha(\lambda_j)|0\rangle_a + \beta(\lambda_j)|1\rangle_a), \tag{6.136}$$

with $|\alpha|^2 + |\beta|^2 = 1$. The transformation is chosen so that

$$\beta(\lambda_j) \sim \frac{1}{\lambda_j},$$

to approximate the desired state (6.130). The transformation is achieved by controlled R_y gates. We will discuss the specific implementation in more detail in the following example. We assume for now that we have implemented $\beta(\lambda_j) = C/\lambda_j$, where C is a constant that has to be chosen in advance. It needs to be as large as possible but smaller than the smallest eigenvalue; otherwise, $\alpha(\lambda_j) = \sqrt{1 - \frac{C^2}{\lambda_j^2}}$ is undefined.

4. *Step 4: Inverse QPE*
Now that we have used the value of λ_j from the first quantum register for the controlled rotation, we can undo the QPE. The inverse QPE leads to $|\lambda_j\rangle \, |u_j\rangle \rightarrow |0\rangle \, |u_j\rangle$, and the resulting state of the system is:

$$|\psi_3\rangle = \sum_j b_j \, |0\rangle \, |u_j\rangle \, (\alpha(\lambda_j)|0\rangle_a + \beta(\lambda_j)|1\rangle_a). \tag{6.137}$$

5. *Step 5: Measurement on the ancilla qubit and postselection*
Now a measurement of the ancilla qubit is performed. Hopefully, the result is 1; otherwise, the computation has failed and needs to be repeated. The reduced ("postselected") state to the ancilla state $|1\rangle_a$ is

$$|\psi_{3,\text{red}}\rangle = \sum_j b_j \beta(\lambda_j) \, |0\rangle \, |u_j\rangle, \tag{6.138}$$

and, after inserting $\beta(\lambda_j) = C/\lambda_j$, we note that the first quantum contains the desired final state $|x\rangle$ (cf. Eq. (6.130)):

$$|\psi_{3,\text{red}}\rangle \sim \sum_j \frac{b_j}{\lambda_j} \, |u_j\rangle. \tag{6.139}$$

This solves the problem. The solution of the linear system of equations has been prepared as the final state $|x\rangle$ in the first quantum register.

Some remarks on the HHL algorithm:

1. To execute the algorithm, the operation e^{iAt} must be realized by gate operations. This has not been discussed in detail in the text because an important area of quantum computing, known as *Hamiltonian simulation*, deals with exactly this problem. Operators of this form describe the time evolution of quantum systems and are therefore intensively studied in the field of quantum simulation.

2. In the beginning, it was mentioned as a condition for the success of the HHL algorithm that the matrix A should be sparse. The algorithm itself works even if this condition is not satisfied. However, the Hamiltonian simulation can be implemented more efficiently for sparse matrices. Thus, the advantage of the quantum algorithm over the most efficient classical algorithms is greater when the matrices under consideration are sparse.

3. The second success condition for the HHL algorithm is that the matrix A must not be ill-conditioned. This can be explained as follows: The success of the algorithm is determined by the probability of finding the value 1 when measuring the ancilla qubit, i. e. by the value of $\beta(\lambda_j) = C/\lambda_j$. Therefore, it is advantageous to choose C as large as possible. On the other hand, C cannot be greater than the smallest eigenvalue. If the matrix A is ill-conditioned, i. e. has a large condition number, the largest and the smallest eigenvalues differ strongly. This reduces the probability of finding the value 1 in the measurement because then C/λ_j is necessarily small for some λ_j. Therefore, the success probability of the HHL algorithm is reduced for ill-conditioned matrices. In general, the eigenvalues of A are not known in advance, but there are ways to estimate them (Courant-Fischer theorem).

Example: Describe the steps of the HHL algorithm explicitly for the following matrix:

$$A = \frac{1}{2}\begin{pmatrix} 3 & 1 \\ 1 & 3 \end{pmatrix}. \tag{6.140}$$

Solution: This matrix was used as the basis in one of the first experimental realizations of the HHL algorithm ([88, 87]). It has the advantage of having "streamlined" eigenvalues that can be encoded in two qubits without rounding error. Its eigenvalues and eigenvectors are

$$\lambda_1 = 2 \quad \text{with } |u_1\rangle = \frac{1}{\sqrt{2}}\begin{pmatrix} 1 \\ 1 \end{pmatrix} = \frac{1}{\sqrt{2}}(|0\rangle + |1\rangle),$$

$$\lambda_2 = 1 \quad \text{with } |u_2\rangle = \frac{1}{\sqrt{2}}\begin{pmatrix} 1 \\ -1 \end{pmatrix} = \frac{1}{\sqrt{2}}(|0\rangle - |1\rangle). \tag{6.141}$$

1. *Step 1: Initialization*
   The first quantum register and the ancilla qubit are set to the state $|0\rangle$; in the second quantum register, the state $|b\rangle = b_1|u_1\rangle + b_2|u_2\rangle$ is prepared. The components b_1 and b_2 are part of the problem specification. We do not need their actual values. The initial state is thus

$$|\psi_0\rangle = |0\rangle\left(b_1|u_1\rangle + b_2|u_2\rangle\right)|0\rangle_a. \tag{6.142}$$

2. *Step 2: Quantum phase estimation*

$$|\psi_1\rangle = \left[\underbrace{|10\rangle}_{=2}\, b_1|u_1\rangle + \underbrace{|01\rangle}_{=1}\, b_2|u_2\rangle\right]|0\rangle_a. \tag{6.143}$$

3. *Step 3: Controlled rotation of the ancilla qubit*
   Now the ancilla qubit has to be rotated so that $\beta(\lambda_j) = C/\lambda_j$. This is done using the R_y gate, which acts on the state $|0\rangle_a$ as follows:

$$R_y(\theta)|0\rangle_a = \cos\frac{\theta}{2}|0\rangle_a + \sin\frac{\theta}{2}|1\rangle_a. \tag{6.144}$$

The angle of rotation of the R_y-grid has to be chosen such that $\sin\frac{\theta}{2} = \frac{C}{\lambda_j}$. Thus, during the calculation, λ_j has to be read from the first quantum register, and the angle θ has to be calculated and passed to the R_y gate. This has to be done coherently, i. e., without any measurement on the first quantum register, using only quantum operations. This is possible in principle, but it is cumbersome. In the experimental implementations of the HHL algorithm, the values of θ were always determined in advance. In our case, they are:

$$\lambda_1 = 2 \Rightarrow \frac{1}{\lambda_1} = \frac{1}{2} \Rightarrow \sin\frac{\theta_1}{2} = \frac{1}{2} \Rightarrow \theta_1 = \frac{\pi}{3},$$

$$\lambda_2 = 1 \Rightarrow \frac{1}{\lambda_2} = 1 \Rightarrow \sin\frac{\theta_2}{2} = 1 \Rightarrow \theta_2 = \pi.$$

So, just for our particular example, the transformation can be done as follows: (a) a CR_y operation controlled by the second qubit of the first register with $\theta = \pi$ and (b) a CR_y operation controlled by the first qubit with $\theta = \pi/3$. If we choose $C = 1$, the resulting state is:

$$|\psi_2\rangle = b_1 |10\rangle |u_1\rangle \left(\cos\frac{\pi}{6}|0\rangle_a + \underbrace{\sin\frac{\pi}{6}}_{=1/2} |1\rangle_a \right) + b_2 |01\rangle |u2\rangle \left(\cos\frac{\pi}{2}|0\rangle_a + \underbrace{\sin\frac{\pi}{2}}_{=1} |1\rangle_a \right).$$

4. *Step 4: Inverse QPE*
   This step "cleans up" the first quantum register by removing the eigenvalues:

$$|\psi_3\rangle = b_1 |00\rangle |u_1\rangle \left(\cos\frac{\pi}{6}|0\rangle_a + \sin\frac{\pi}{6}|1\rangle_a \right) + b_2 |00\rangle |u2\rangle \left(\cos\frac{\pi}{2}|0\rangle_a + \sin\frac{\pi}{2}|1\rangle_a \right).$$

5. *Step 5: Measurement at the ancilla qubit and postselection*
   In the case where the measurement at the ancilla qubit yields 1, the postselected state of the remaining system is:

$$|\psi_{3,red}\rangle = |00\rangle \left(\frac{b_1}{2} |u_1\rangle + \frac{b_2}{1} |u_2\rangle \right),$$

which also be written as $|00\rangle |x\rangle$. Thus, the algorithm has prepared the desired state $|x\rangle$ in the second quantum register.

6.9 Quantum error correction

Each of the quantum algorithms discussed so far – Grover, Shor, and HHL – solves a specific class of problems, each with its own basic idea, using the principles of quantum physics in different ways. They have in common that they assume ideal conditions and fully exploit the possibilities of quantum physics, such as superposition and entanglement. However, these ideal conditions do not correspond to today's reality. Current quantum computers are still highly susceptible to decoherence (commonly referred to as noise). In particular, entangled states of multiple qubits are fragile and difficult to realize in practice.

The current stage in the development of quantum computers is called the *NISQ era*, the era of *"Noisy Intermediate-Scale Quantum Computers"*. None of the quantum algorithms described above can be implemented on current quantum computers for practical computation. Even if the nominal number of qubits were available to implement the Grover or Shor algorithm, the results would be indistinguishable from pure noise.

Two approaches have been proposed to deal with the noise problem: (1) the *quantum error correction* approach, in which multiple physical qubits are combined into a single logical qubit in order to store quantum information in a more robust manner, and (2) the *NISQ algorithms* approach, which identifies algorithms that are less susceptible to decoherence. We will discuss an implementation example of each below. While NISQ algorithms are seen as a transitional technology that should be replaced by more powerful quantum algorithms as better quantum computers become available, quantum error correction will also, in the future, be a part of any quantum computer because the suppression of decoherence will always be necessary.

Prerequisites for quantum error correction

We consider qubits in a quantum computer, which, due to their physical implementation, are subject to a number of decoherence mechanisms (sources of noise). As in classical information processing, errors can be corrected by redundancy. The quantum information is encoded in a larger number of qubits than actually needed. During the calculation, various tests are performed to detect possible errors and their nature. If an error is found, it may be possible to correct it. For classical bits, this is a routine procedure, but for qubits, it is much more difficult. Remember: the state cannot be read out by measurement without destroying superposition states, and because of the no-cloning theorem, you cannot make a copy of an unknown quantum state. It is necessary to use more subtle methods.

To successfully perform quantum error correction, a model of the possible errors is needed. The most general possible error in qubit states can be modeled as a combination of unwanted interactions of the qubits with the environment (as in the discussion of decoherence in Section 3.10) and unwanted internal state changes of the qubits. Each quantum error correction method deals with a specific, hopefully large, subset of this general case and tries to provide efficient methods for correcting the particular type of error within a concrete error model.

In the following, we consider a very simple example, which has been first discussed by Steane [89]. The error model has only one type of error: With a certain probability p, the value of exactly one qubit is flipped, i. e. $|0\rangle$ becomes $|1\rangle$ and vice versa (also in superposition states). Thus, the error has the same effect as using the Pauli-X gate. Such a bit flip is by no means the most common type of error that occurs in reality (that would be uncontrolled phase changes, loss of coherence, and the destruction of superposition and entanglement), but the procedure to correct it is particularly transparent and easy to understand.

Encoding and error detection

The logical state to be encoded is assumed to be the general superposition state of a single qubit:

$$|\psi\rangle = a\,|0\rangle + b\,|1\rangle . \tag{6.145}$$

This logical state is encoded in three physical qubits q_0, q_1, q_2, which contain the information redundantly:

$$|\psi\rangle = a\,|000\rangle + b\,|111\rangle . \tag{6.146}$$

The preparation of such a state does not violate the no-cloning theorem because no previously unknown state is cloned, but a previously determined state is prepared.

We now let this system of three qubits interact with the noise source. With probability $1 - p$, no error occurs and the system remains in state (6.146). With probability p, one of the three qubits is flipped. In reality, there would also be a (smaller) probability that two or all three qubits would be affected. In our model, however, we ignore this possibility. The state of the system after exposure to the noise source is thus

$$|\psi\rangle = a\,|000\rangle + b\,|111\rangle \quad \text{or}$$
$$|\psi\rangle = a\,|100\rangle + b\,|011\rangle \quad \text{or}$$
$$|\psi\rangle = a\,|010\rangle + b\,|101\rangle \quad \text{or}$$
$$|\psi\rangle = a\,|001\rangle + b\,|110\rangle . \tag{6.147}$$

The "or" should imply that the state of the system is a superposition of all four possibilities. However, since we will end up performing a measurement with appropriate state reduction, the superposition is not relevant here, and we can treat the four possibilities as classical alternatives.

In order to detect whether a qubit flip has occurred (and if so, at which of the three qubits), we employ two additional ancilla qubits a_0 and a_1, which are set to the state $|00\rangle$ initially. We perform pairwise comparisons of the qubits q_0 and q_1 and q_0 and q_2 and write the results into the ancilla qubits. The aim is to determine which of the qubits deviates from the other two by performing the two pairwise comparisons and then to correct the error specifically.

We have already seen in the example task on p. 159 that two qubits can be compared using two CNOT gates (Fig. 6.15). If they match, the ancilla qubit will afterwards be in the

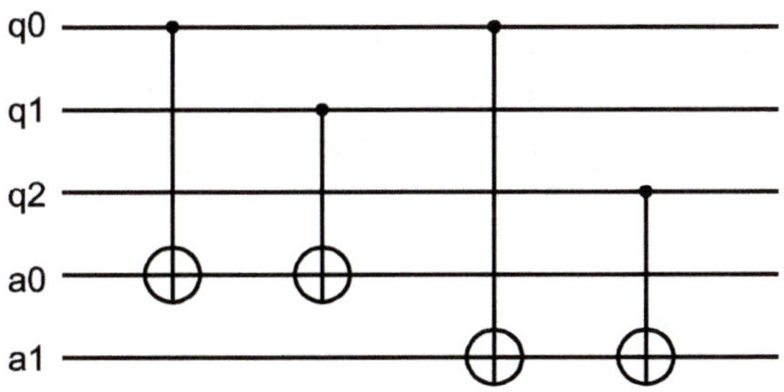

Figure 6.31: Detection of qubit flip errors using four CNOT gates.

state $|0\rangle$, otherwise in $|1\rangle$. Thus, for the two pairwise comparisons, a total of four CNOT operations have to be performed, as shown in Fig. 6.31. The state of the whole system is then:

$$
\begin{aligned}
|\psi\rangle &= (a\,|000\rangle + b\,|111\rangle)\,|00\rangle \quad \text{or} \\
|\psi\rangle &= (a\,|100\rangle + b\,|011\rangle)\,|11\rangle \quad \text{or} \\
|\psi\rangle &= (a\,|010\rangle + b\,|101\rangle)\,|10\rangle \quad \text{or} \\
|\psi\rangle &= (a\,|001\rangle + b\,|110\rangle)\,|01\rangle,
\end{aligned}
\tag{6.148}
$$

where the states on the right represent the two ancilla qubits.

Example: Verify the result of Eq. (6.148) for the state in the third line.

Solution: The initial state is:

$$
|\psi_0\rangle = a\,|010\rangle\,|00\rangle + b\,|101\rangle\,|00\rangle .
\tag{6.149}
$$

The four CNOT operations are now applied in sequence. As a reminder, the CNOT gate inverts the target qubit for those state components where the control qubit is 1. This results in the following:

$$
\begin{aligned}
\text{CNOT}_{(q_0,a_0)}: \quad & |\psi_1\rangle = a\,|\underline{0}10\rangle\,|\underline{0}0\rangle + b\,|\underline{1}01\rangle\,|\underline{1}0\rangle , & (6.150) \\
\text{CNOT}_{(q_1,a_0)}: \quad & |\psi_2\rangle = a\,|0\underline{1}0\rangle\,|\underline{1}0\rangle + b\,|1\underline{0}1\rangle\,|\underline{1}0\rangle , & (6.151) \\
\text{CNOT}_{(q_0,a_1)}: \quad & |\psi_3\rangle = a\,|\underline{0}10\rangle\,|1\underline{0}\rangle + b\,|\underline{1}01\rangle\,|1\underline{1}\rangle , & (6.152) \\
\text{CNOT}_{(q_2,a_1)}: \quad & |\psi_4\rangle = a\,|01\underline{0}\rangle\,|1\underline{0}\rangle + b\,|10\underline{1}\rangle\,|1\underline{0}\rangle . & (6.153)
\end{aligned}
$$

Thus, after applying the four CNOT gates, the final state is the one as specified in the third line of Eq. (6.148).

The two pairwise comparisons have achieved their goal: In Eq. (6.148), each of the four possible qubit states is associated with a unique state of the two ancilla qubits. Now, a measurement of the two ancilla qubits is performed, and one of the four alternatives is obtained as a result. For example, if the measurement yields the value "10", this indicates that the second qubit q_1 is affected by an error (third line in Eq. (6.148)). Without measuring this qubit itself and without destroying the superposition state, one can now selectively apply a Pauli-X gate and thereby correct the error. Qubits that – like the two ancilla qubits here – indicate the nature and location of an error are called *syndrome qubits* in quantum error correction.

Advanced quantum error correction techniques

Quantum error correction is a well-developed field of research, mostly characterized by a high mathematical level because of the need to quantify errors. Therefore, we will only give an overview of the basic features of the various error correction methods. From our simple example, we can already see some of them:

1. In order to detect and correct errors, the state space is extended by combining several physical qubits into one logical qubit. The high dimensionality of quantum mechanical state spaces is used to encode information as robustly and redundantly as possible.

2. For error detection, measurements are made that do not destroy the crucial superpositions. Such measurements are called *stabilizer measurements* because they leave the logical qubit intact and help to detect errors. Stabilizer measurements can usually be written as products of Pauli operators acting on different qubits with eigenvalues ± 1 (error/no error). Each quantum error correction method involves a whole set of stabilizer measurements. All of them must commute with each other so as not to interfere with each other. Their construction is usually done using group theory methods.

3. The results of the stabilizer measurements provide the syndrome, which gives clues to the nature and location of the error that has occurred. Actions can then be taken to correct the error.

In many practical quantum computing realizations, a qubit can interact with only a few nearest neighbors. This is especially true for superconducting qubits, where each qubit has a fixed location on the chip and hardwired connections to its neighbors (see Fig. 1.2 on p. 4). Therefore, logical qubit implementations based only on nearest-neighbor interaction are preferred. Commonly used methods are the *surface code* or *color codes*, which employ different geometric arrangements of qubits on an abstract lattice.

A useful implementation of quantum algorithms becomes possible when it is possible to correct errors faster than they occur. The *quantum threshold theorem* states that this is possible in principle with quantum error correction. A general goal is to achieve an error rate of less than 1 % for each operation. Achieving this goal poses different challenges for different hardware implementations of quantum computers: superconducting qubits, for example, are inherently more noisy than trapped-ion-based qubits.

However, even below the threshold, any noise reduction that can be achieved on the hardware side is worthwhile: The higher the error rate, the more physical qubits are needed to realize an error-corrected logical qubit. Their number increases very rapidly with the error rate and can easily reach several hundred or thousand. Because of this strong dependence on the error rate, it is problematic to compare different quantum computer realizations solely on the basis of the number of physical qubits.

6.10 NISQ algorithms

The current generation of quantum computers does not yet have full quantum error correction. The "classical" quantum algorithms cannot be run on them in a useful way. Therefore, *NISQ algorithms* are currently being explored. Their output is insensitive to a certain amount of noise [90]. NISQ algorithms are designed so that the errors caused by noise do not completely prevent the operation of the algorithm but only manifest themselves in tolerable inaccuracies in the result. Typical NISQ algorithms take a hybrid approach, coupling a quantum computer with a powerful classical computer. The quantum computer executes the part of the algorithm that makes use of the large state space spanned by the qubits, while the rest of the computation is performed by the classical computer.

NISQ optimization algorithms

The most common NISQ algorithms are optimization algorithms. They are based on the long-established variational principles of atomic physics and quantum chemistry. To find the stable state of a (possibly complex) molecule, one makes an approximation for the electron distribution depending on some parameters and varies these parameters until minimum energy is found. The result is an approximation for the ground state, the stable configuration of the molecule.

A direct implementation of this idea is the *variational quantum eigensolver* (VQE): the step that is "computationally expensive" for a classical computer is the calculation of the energy for each state. This is done by a quantum computer, while the optimization itself is delegated to a classical computer using one of the established optimization algorithms [91]. The quantum algorithm receives as input the system parameters for which the energy value is to be computed. The classical algorithm changes the parameters according to these results to find a minimum of the energy.

It is straightforward to apply this method to other optimization problems. The energy is replaced by a cost function to be minimized, usually under some constraints. The task in implementing the quantum algorithm is to map the problem to an interaction of qubits. Some standard methods have been established, mostly using notions from mathematical graph theory, because many practical optimization problems can be represented in this way.

Tail assignment problem

A specific example from the field of aviation illustrates the nature of the problems that can be solved. An airline has to serve connections between different airports as efficiently and cheaply as possible using a given

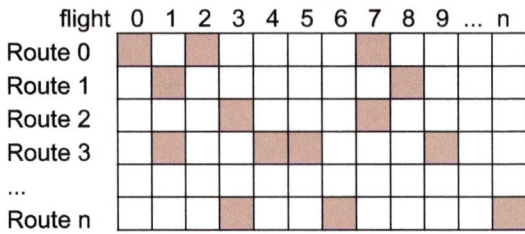

Figure 6.32: Tail assignment problem: Different flights have to be combined into routes as efficiently as possible. Pink cells correspond to a 1; white cells correspond to a 0.

number of planes. The problem is illustrated in Fig. 6.32: The columns of the table contain the connections that have to be served. The rows represent different possible flight routes. Each route consists of several individual flights, which have to be chosen in such a way that the route is continuous. A pink cell indicates that the corresponding flight is included in the route. The goal is to make a selection from the possible routes in which each connection is flown exactly once. Therefore, the algorithm should select rows from the table in such a way that each column contains exactly one colored cell. If there are several possibilities, the least expensive should be chosen. This problem is called *tail assignment problem;* the name comes from the number plate on the tail of an airplane, which is used to identify and plan flight routes. The problem was formulated for the solution with quantum algorithms in the following way [92, 93]. The cost function to be minimized is:

$$\sum_r c_r x_r, \tag{6.154}$$

where the sum covers all possible routes (i. e., rows of the table in Fig. 6.32). The variable x_r is equal to 1 if the route r is one of the selected ones, and 0 otherwise. c_r indicates the cost of the route r. In addition, the constraints

$$\sum_r A_{fr} x_r = 1 \quad \text{for all } f, \tag{6.155}$$

have to be satisfied, which ensure that all flights f are represented exactly once. The matrix A_{fr} corresponds to the table shown in Fig. 6.32, where pink cells correspond to 1 and white ones to 0.

In [92, 93], the tail assignment problem was tackled using the *Quantum Approximate Optimization Algorithm* (QAOA), the most widely used NISQ optimization algorithm [94]. For this purpose, the cost function (6.154) and the constraints (6.155) have to be mapped onto a system of qubits whose interaction is described by the Hamiltonian

$$H_C = \sum_{i<j} J_{ij} Z_i Z_j + \sum_i h_i Z_i. \tag{6.156}$$

Here, Z_i are the Pauli-Z matrices, and J_{ij} and h_i are the strength of the interaction. The interaction determined by Eq. (6.156) is also called the *Ising model.*

The idea behind the algorithm is to minimize the expectation value of the "cost Hamiltonian" $\langle\psi(\beta,\gamma)|H_C|\psi(\beta,\gamma)\rangle$, which describes the cost function. The state $|\psi(\beta,\gamma)\rangle$

is obtained from the initial state of the qubits by multiple applications of the operators $U(\gamma) = e^{-i\gamma H_C}$ and $U(\beta) = e^{-i\beta H_M}$, where the operator H_M does not commute with H_C and is called a mixing Hamiltonian. In a sense, it drives the state through different regions of the state space. The qubits in the state that has been prepared in this way are now measured repeatedly, and from the results, the expectation value $\langle \psi(\beta, \gamma)|H_C|\psi(\beta, \gamma)\rangle$ is calculated. This is the part of the calculation that is difficult for a classical computer.

The actual optimization by systematic variation of the parameters β and γ is done by a classical algorithm. It determines new values for β and γ, which are then passed to the quantum algorithm in a new iteration step. This continues until a satisfactory approximation to the desired minimum is obtained. The sequence of qubits in this state then contains the desired solution – in the case of the tail assignment problem, the selected and unselected routes encoded as 0 and 1.

An important step towards demonstrating that such approach works for practically relevant problems was made by researchers at IBM and the University of California at Berkeley in 2023 [95]. They simulated an Ising model of the form (6.156) in three different ways: (1) using a quantum computer with 127 qubits, (2) with modern approximation methods ("tensor networks") on a classical supercomputer, and (3) with exact "brute force" methods on another supercomputer. The third method was only feasible up to a certain problem size. As the complexity of the problem increased, the solution provided by the quantum computer differed from the result of the classical approximation method. It turned out that the quantum result was a more accurate match to the exact solution than the classical approximation. This may indicate that the quantum computer provided more reliable results even in those parameter ranges where comparison with the exact solution was no longer possible.

The key to this success, which gives hope that quantum computers can be useful even without sophisticated error correction, was the careful control of qubit errors. The size of the errors could be modeled in detail and extrapolated to zero. This provided an estimate of the exact result. Such methods, where errors are not corrected but modeled, and their impact on the final result is estimated, are called *Quantum Error Mitigation*. They are an important complement to quantum error correction.

A Common quantum gates

identity in —[𝟙]— out

in	out
$|0\rangle$	$|0\rangle$
$|1\rangle$	$|1\rangle$

Pauli X in —[X]— out

in	out
$|0\rangle$	$|1\rangle$
$|1\rangle$	$|0\rangle$

Pauli Z in —[Z]— out

in	out
$|0\rangle$	$|0\rangle$
$|1\rangle$	$-|1\rangle$

Hadamard in —[H]— out

in	out
$|0\rangle$	$\frac{1}{\sqrt{2}}(|0\rangle + |1\rangle)$
$|1\rangle$	$\frac{1}{\sqrt{2}}(|0\rangle - |1\rangle)$

CNOT

in1 ——●—— out1
in2 ——⊕—— out2

in1	in2	out1	out2
$|0\rangle$	$|0\rangle$	$|0\rangle$	$|0\rangle$
$|0\rangle$	$|1\rangle$	$|0\rangle$	$|1\rangle$
$|1\rangle$	$|0\rangle$	$|1\rangle$	$|1\rangle$
$|1\rangle$	$|1\rangle$	$|1\rangle$	$|0\rangle$

Toffoli

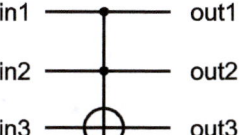

in1 ——●—— out1
in2 ——●—— out2
in3 ——⊕—— out3

in1	in2	in3	out1	out2	out3
$|0\rangle$	$|0\rangle$	$|0\rangle$	$|0\rangle$	$|0\rangle$	$|0\rangle$
$|0\rangle$	$|0\rangle$	$|1\rangle$	$|0\rangle$	$|0\rangle$	$|1\rangle$
$|0\rangle$	$|1\rangle$	$|0\rangle$	$|0\rangle$	$|1\rangle$	$|0\rangle$
$|0\rangle$	$|1\rangle$	$|1\rangle$	$|0\rangle$	$|1\rangle$	$|1\rangle$
$|1\rangle$	$|0\rangle$	$|0\rangle$	$|1\rangle$	$|0\rangle$	$|0\rangle$
$|1\rangle$	$|0\rangle$	$|1\rangle$	$|1\rangle$	$|0\rangle$	$|1\rangle$
$|1\rangle$	$|1\rangle$	$|0\rangle$	$|1\rangle$	$|1\rangle$	$|1\rangle$
$|1\rangle$	$|1\rangle$	$|1\rangle$	$|1\rangle$	$|1\rangle$	$|0\rangle$

https://doi.org/10.1515/9783110717457-007

CCZ

in1 ———•——— out1

in2 ———•——— out2

in3 ——[Z]—— out3

in1	in2	in3	out1	out2	out3
$|0\rangle$	$|0\rangle$	$|0\rangle$	$|0\rangle$	$|0\rangle$	$|0\rangle$
$|0\rangle$	$|0\rangle$	$|1\rangle$	$|0\rangle$	$|0\rangle$	$|1\rangle$
$|0\rangle$	$|1\rangle$	$|0\rangle$	$|0\rangle$	$|1\rangle$	$|0\rangle$
$|0\rangle$	$|1\rangle$	$|1\rangle$	$|0\rangle$	$|1\rangle$	$|1\rangle$
$|1\rangle$	$|0\rangle$	$|0\rangle$	$|1\rangle$	$|0\rangle$	$|0\rangle$
$|1\rangle$	$|0\rangle$	$|1\rangle$	$|1\rangle$	$|0\rangle$	$|1\rangle$
$|1\rangle$	$|1\rangle$	$|0\rangle$	$|1\rangle$	$|1\rangle$	$|0\rangle$
$|1\rangle$	$|1\rangle$	$|1\rangle$	$|1\rangle$	$|1\rangle$	$-|1\rangle$

phase gate R_k

in ——[R_k]—— out

in	out
$|0\rangle$	$|0\rangle$
$|1\rangle$	$\exp(\frac{2\pi i}{2^k})\,|1\rangle$

CR_k

in1 ———•——— out1

in2 ——[R_k]—— out2

in1	in2	out1	out2
$|0\rangle$	$|0\rangle$	$|0\rangle$	$|0\rangle$
$|0\rangle$	$|1\rangle$	$|0\rangle$	$|1\rangle$
$|1\rangle$	$|0\rangle$	$|1\rangle$	$|0\rangle$
$|1\rangle$	$|1\rangle$	$|1\rangle$	$\exp(\frac{2\pi i}{2^k})\,|1\rangle$

Image credits

Christian Kurtsiefer: Fig. 3.2
David Nadlinger, University of Oxford: Fig. 2.4
PTB: Fig. 1.4 (research group Christian Ospelkaus), 1.7, 1.10 (Thomas Middelmann)
Pixabay: Fig. 3.14 (yeTis); 6.32 (Andrew Sitnikov)
QZabre: Fig. 1.9
Tobias Reinsch, Stuttgart University: Fig. 1.8

Wikimedia Commons: Fig. 2.1 (Tigerzeng); 2.3 (Marcin Wichary); 1.5 (Jpagett); 1.6 (Mwjohnson0); 1.11 (Radovan Blazek); 3.12 left (Zaereth); 3.12 right (Peeter Piksarv); 5.6 (Hubert Berberich)

https://doi.org/10.1515/9783110717457-008

Bibliography

[1] Binosi D et al. EuroQCS white paper: European quantum computing & simulation infrastructure. Techn. Rep. Quantum Flagship; 2022.

[2] Feynman RP. Simulating physics with computers. Int J Theor Phys. 1982; 21(6):467–488.

[3] Bloch I, Dalibard J, Nascimbène S. Quantum simulations with ultracold quantum gases. Nat Phys. 2012; 8:267–276.

[4] Berger C et al. Quantum technologies for climate change: Preliminary assessment. 2021. arXiv:2107.05362.

[5] Kagermann H et al. The innovation potential of second-generation quantum technologies. acatech IMPULSE. Munich; 2020.

[6] Battersby S, Ed. The quantum age: Technological opportunities. Government Office for Science, London; 2016.

[7] Bennett CH et al. Teleporting an unknown quantum state via dual classical and Einstein–Podolsky–Rosen channels. Phys Rev Lett. 1993; 70(13):18951899.

[8] Boschi D et al. Experimental realization of teleporting an unknown pure quantum state via dual classical and Einstein–Podolsky–Rosen channels. Phys Rev Lett. 1998; 80(6):1121–1125.

[9] Bouwmeester D et al. Experimental quantum teleportation. Nature. 1997; 390(6660):575–579.

[10] Nagourney W, Sandberg J, Dehmelt H. Shelved optical electron amplifier: Observation of quantum jumps. Phys Rev Lett. 1986; 56(26):2797–2799.

[11] Monroe C et al. Demonstration of a fundamental quantum logic gate. Phys Rev Lett. 1995; 75(25):4714–4717.

[12] Feynman RP. The Feynman lectures on Physics, Vol. I, Ch. 37: Quantum behavior. 1962. https://www.feynmanlectures.caltech.edu/I_37.html#Ch1-audio.

[13] Carnal O, Mlynek J. Young's double-slit experiment with atoms: A simple atom interferometer. Phys Rev Lett. 1991; 66(21):2689–2692.

[14] Küblbeck J, Müller R. Die Wesenszüge der Quantenphysik: Modelle, Bilder und Experimente. Praxis-Schriftenreihe Abteilung Physik 60. Köln: Aulis-Verl. Deubner; 2002.

[15] Müller R, Mishina O. Milq—quantum physics in secondary school. In: Teaching-learning contemporary physics: from research to practice. Cham: Springer International Publishing; 2021. P. 33–45.

[16] Audretsch J. Die sonderbare Welt der Quanten: eine Einführung. München: Beck; 2008.

[17] Pfau T et al. Loss of spatial coherence by a single spontaneous emission. Phys Rev Lett. 1994; 73(9):1223–1226.

[18] Englert B-G. Fringe visibility and which-way information: An inequality. Phys Rev Lett. 1996; 77(11):2154–2157.

[19] Grangier P, Roger G, Aspect A. Experimental evidence for a photon anticorrelation effect on a beam splitter: a new light on single-photon interferences. Europhys Lett. 1986; 1(4):173–179.

[20] Press WH, Ed. Numerical recipes in C: The art of scientific computing. 2nd ed. Cambridge; New York: Cambridge University Press; 1992.

[21] Federal Office for Information Security (BSI), Ed. Quantum-safe cryptography. Bonn; 2021. BSI-Bro21/01.

[22] Cleve R et al. Quantum algorithms revisited. Proc R Soc, Math Phys Eng Sci. 1998; 454(1969):339–354.

[23] Pade J. Quantum mechanics for pedestrians 1: Fundamentals. Cham, Switzerland: Springer; 2018.

[24] Holbrow CH, Galvez E, Parks ME. Photon quantum mechanics and beam splitters. Am J Phys. 2002; 70(3):260–265.

[25] Henault F. Quantum physics and the beam splitter mystery. In: The nature of light: What are photons? VI. Vol. 9570. SPIE; 2015; 199–213.

[26] Werner R. The uncertainty relation for joint measurement of position and momentum. Quantum Inf Comput. 2004; 4:546–562.

https://doi.org/10.1515/9783110717457-009

[27] Ozawa M. Uncertainty relations for noise and disturbance in generalized quantum measurements. Ann Phys. 2004; 311(2):350–416.

[28] Busch P, Lahti und P, Werner RF. Proof of Heisenberg's error–disturbance relation. Phys Rev Lett. 2013; 111(16):160405.

[29] Schrödinger E. Die gegenwärtige Situation in der Quantenmechanik. Naturwissenschaften. 1935; 23(48):807–812.

[30] Fein YY et al. Quantum superposition of molecules beyond 25 kDa. Nat Phys. 2019; 15(12):1242–1245.

[31] Zeh HD. On the interpretation of measurement in quantum theory. Found Phys. 1970; 1(1):69–76.

[32] Zurek WH. Decoherence and the transition from quantum to classical. Phys Today. 1991; 44(10):36–44.

[33] Englert B-G. On quantum theory. Eur Phys J D. 2013; 67(11):238.

[34] Einstein A, Born H, Born M. Briefwechsel 1916–1955. Reinbek bei Hamburg: Rowohlt; 1972.

[35] Bertlmann RA. Magic moments with John Bell. Phys Today. 2015; 68(7):40–45.

[36] Clauser JF, Horne MA. Experimental consequences of objective local theories. Phys Rev D. 1974; 10(2):526–535.

[37] Kochen und S, Specker EP. The problem of hidden variables in quantum mechanics. J Math Mech. 1967; 17(1):59–87.

[38] Aspect A, Grangier P, Roger G. Experimental realization of Einstein–Podolsky–Rosen–Bohm Gedankenexperiment: A new violation of Bell's inequalities. Phys Rev Lett. 1982; 49(2):91–94.

[39] Kwiat PG et al. New high-intensity source of polarization-entangled photon pairs. Phys Rev Lett. 1995; 75(24):4337–4341.

[40] Schirhagl R et al. Nitrogen-vacancy centers in diamond: Nanoscale sensors for physics and biology. Annu Rev Phys Chem. 2014; 65(1):83–105.

[41] Waasem N, Fedder H, Maletinsky P. Technology leaps in quantum sensing: Advances in nano magnetometry using tailored electronics and fast-switchable lasers. Photonics Views. 2021; 18(4):36–39.

[42] Zhang J, Hegde SS, Suter D. Efficient implementation of a quantum algorithm in a single nitrogen-vacancy center of diamond. Phys Rev Lett. 2020; 125(3):030501.

[43] Elitzur AC, Vaidman L. Quantum mechanical interaction-free measurements. Found Phys. 1993; 23:987–997.

[44] Kwiat P et al. Interaction-free measurement. Phys Rev Lett. 1995; 74(24):4763–4766.

[45] Gilaberte Basset M et al. Perspectives for applications of quantum imaging. Laser Photonics Rev. 2019; 13(10):1900097.

[46] Padgett und MJ, Boyd RW. An introduction to ghost imaging: Quantum and classical. Philos Trans R Soc, Math Phys Eng Sci. 2017; 375(2099):20160233.

[47] Pittman TB et al. Optical imaging by means of two-photon quantum entanglement. Phys Rev A. 1995; 52(5):R3429–R3432.

[48] Aspden RS et al. Photon-sparse microscopy: visible light imaging using infrared illumination. Optica. 2015; 2(12):1049–1052.

[49] Bennink RS, Bentley und SJ, Boyd RW. 'Two-photon' coincidence imaging with a classical source. Phys Rev Lett. 2002; 89(11):113601.

[50] Lemos GB et al. Quantum imaging with undetected photons. Nature. 2014; 512(7515):409–412.

[51] Dowling JP. Quantum optical metrology – the lowdown on high-N00N states. Contemp Phys. 2008; 49(2):125–143.

[52] Giovannetti V, Lloyd S, Maccone L. Advances in quantum metrology. Nat Photonics. 2011; 5(4):222–229.

[53] Wineland DJ et al. Quantum computers and atomic clocks. In: Frequency standards and metrology. World Scientific; 2002; 361–368.

[54] Schmidt PO et al. Spectroscopy using quantum logic. Science. 2005; 309(5735):749–752.

[55] Schmidt PO et al. Quantum logic for precision spectroscopy. Special Issue/PTB- Mitteilungen. 2009; 119(2):54–59.

[56] Brewer SM et al. ^{27}Al$^+$ quantum-logic clock with a systematic uncertainty below 10^{-18}. Phys Rev Lett. 2019; 123(3):033201.

[57] Werner RF. Quantum information theory – An invitation. In: von Alber G et al., Ed. Quantum information: An introduction to basic theoretical concepts and experiments. Springer tracts in modern physics. Berlin, Heidelberg: Springer; 2001; 14–57.

[58] Franz T, Werner RF. Unmögliche Maschinen in der Quantenphysik. Prax Nat.wiss, Phys. 2016; 65(1):39–43.

[59] Wootters WK, Zurek WH. A single quantum cannot be cloned. Nature. 1982; 299(5886):802–803.

[60] Bennett CH, Brassard G. Quantum cryptography: Public key distribution and coin tossing. Theor Comput Sci. 2014; 560:7–11.

[61] Ekert AK. Quantum cryptography based on Bell's theorem. Phys Rev Lett. 1991; 67(6):661–663.

[62] Gisin N, Thew R. Quantum communication. Nat Photonics. 2007; 1(3):165–171.

[63] Hwang W-Y. Quantum key distribution with high loss: Toward global secure communication. Phys Rev Lett. 2003; 91(5):057901.

[64] Bennett CH et al. Experimental quantum cryptography. J Cryptol. 1992; 5(1):3–28.

[65] Schuck C et al. Complete deterministic linear optics bell state analysis. Phys Rev Lett. 2006; 96(19):190501.

[66] Boaron A et al. Secure quantum key distribution over 421 km of optical fiber. Phys Rev Lett. 2018; 121(19):190502.

[67] Yin J et al. Satellite-based entanglement distribution over 1200 kilometers. Science. 2017; 356(6343):1140–1144.

[68] Pirandola S et al. Fundamental limits of repeaterless quantum communications. Nat Commun. 2017; 8(1):15043.

[69] Briegel H-J et al. Quantum repeaters: The role of imperfect local operations in quantum communication. Phys Rev Lett. 1998; 81(26):5932–5935.

[70] van Leent T et al. Entangling single atoms over 33 km telecom fibre. Nature. 2022; 607(7917):69–73.

[71] Krutyanskiy V et al. Telecom-wavelength quantum repeater node based on a trapped-ion processor. Phys Rev Lett. 2023; 130(21):213601.

[72] Bennett CH et al. Purification of noisy entanglement and faithful teleportation via noisy channels. Phys Rev Lett. 1996; 76(5):722–725.

[73] DiVincenzo DP. The physical implementation of quantum computation. Fortschr Phys. 2000; 48(9–11):771–783.

[74] Meyer DA. Quantum strategies. Phys Rev Lett. 1999; 82(5):10521055.

[75] Grover LK. A fast quantum mechanical algorithm for database search. In: Proceedings of the twenty-eighth annual ACM Symposium on Theory of Computing, STOC '96. New York, NY, USA: Association for Computing Machinery; 1996; 212–219.

[76] Nielsen MA, Chuang IL. Quantum computation and quantum information: 10th anniversary edition. Cambridge University Press; 2010.

[77] Grover LK. From Schrödinger's equation to the quantum search algorithm. Pramana. 2001; 56(2–3):333–348.

[78] Bennett CH et al. Strengths and weaknesses of quantum computing. SIAM J Comput. 1997; 26(5):1510–1523.

[79] Coppersmith D. An approximate Fourier transform useful in quantum factoring. 2002. arXiv:quant-ph/0201067.

[80] Shor PW. Algorithms for quantum computation: Discrete logarithms and factoring. In: Proceedings 35th annual symposium on foundations of computer science. 1994; 124134.

[81] Mermin ND. What has quantum mechanics to do with factoring? Phys Today. 2007; 60(4):8–9.

[82] Shor PW. Polynomial-time algorithms for prime factorization and discrete logarithms on a quantum computer. SIAM Rev. 1999; 41(2):303–332.

[83] Kitaev AY. Quantum measurements and the Abelian stabilizer problem. Electron Colloq Comput Complex. 1996; 3.

[84] Harrow AW, Hassidim A, Lloyd S. Quantum algorithm for linear systems of equations. Phys Rev Lett. 2009; 103(15):150502.

[85] Dervovic D et al. Quantum linear systems algorithms: A primer. 2018. arXiv:1802.08227.

[86] Duan B et al. A survey on HHL algorithm: From theory to application in quantum machine learning. Phys Lett A. 2020; 384(24):126595.

[87] Barz S et al. A two-qubit photonic quantum processor and its application to solving systems of linear equations. Sci Rep. 2014; 4(1):6115.

[88] Pan J et al. Experimental realization of quantum algorithm for solving linear systems of equations. Phys Rev A. 2014; 89(2):022313.

[89] Steane AM. Quantum computing and error correction. 2003. arXiv:quant-ph/0304016.

[90] Bharti K et al. Noisy intermediate-scale quantum algorithms. Rev Mod Phys. 2022; 94(1):015004.

[91] Peruzzo A et al. A variational eigenvalue solver on a photonic quantum processor. Nat Commun. 2014; 5(1):4213.

[92] Vikstal P et al. Applying the quantum approximate optimization algorithm to the tail-assignment problem. Phys Rev Appl. 2020; 14(3):034009.

[93] Willsch M et al. Benchmarking the quantum approximate optimization algorithm. Quantum Inf Process. 2020; 19(7):197.

[94] Farhi E, Goldstone J, Gutmann S. A quantum approximate optimization algorithm. 2014. arXiv:1411.4028.

[95] Kim Y et al. Evidence for the utility of quantum computing before fault tolerance. Nature. 2023; 618:500–505.

Index

https://doi.org/10.1515/9783110717457-010